geothermics

geothermics

JEAN GOGUEL

Professeur à l'Ecole Nationale Superièure des Mines de Paris

English language edition edited by
Professor Sydney P. Clark, Jr.
Sidney J. Weinberg Professor of Geophysics
Yale University

Translated by
Alan Rite
Department of Geology and Geophysics
Yale University

McGraw-Hill Book Company
*New York   St. Louis   San Francisco   Auckland   Düsseldorf
Johannesburg   Kuala Lumpur   London   Mexico   Montreal
New Delhi   Panama   Paris   São Paulo   Singapore   Sydney
Tokyo   Toronto*

This book was set in Century Expanded by Textbook Services, Inc.
The editors were Bradford Bayne, Patrick A. Clifford, and
Richard S. Laufer;
the designer was Barbara Ellwood;
the production supervisor was Judi Allen.
Kingsport Press, Inc., was printer and binder.

Library of Congress Cataloging in Publication Data

Goguel, Jean.
    Geothermics.

    (McGraw-Hill international series in the earth and planetary sciences)
    Translation of La géothermie.
    1. Geothermal resources. 2. Water, Underground.
3. Earth temperature. I. Title.
GB1199.5.G6313   553   75-28177
ISBN 0-07-023518-X

geothermics

1 2 3 4 5 6 7 8 9 0 KPKP 7 8 3 2 1 0 9 8 7 6

Originally published in French under the title of LA GEOTHERMIE by
Doin Editeurs, Paris, France. Copyright © 1975 by Doin Editeurs.

contents

editor's introduction

This volume is the English edition of "La Géothermie," which was originally published in Paris by Doin Editeurs. Most of the book is devoted to a discussion, beginning with first principles, of the thermal regime near the surface of the earth, with particular reference to the interaction between the temperature field and circulating groundwater. The results are applied to the problem of extracting usable heat and power from geothermal sources. The operation of generating plants using both wet and dry steam are extensively described, and methods developed by the author to estimate the power output and useful lifetime to be expected from a given geothermal area are also discussed. One chapter is devoted to the economic and political constraints that must be satisfied if geothermal energy is to become a reality.

The final two chapters are devoted to topics of interest to students of earth science. One deals with the relation between geothermics and certain selected geological phenomena. These include the thermal effects of igneous intrusions and extrusions, heat produced by deformation and faulting, thermal stresses in rocks, and the problem of the source of the heat producing metamorphism. The final chapter addresses itself to the thermal state of the earth as a whole, discussing the cooling of the earth, the radioactivity of earth materials (including the important thermal effects of radioactive decay), the thermal energy budget of the earth, and the author's views on convective circulation in the mantle.

The author is presently Ingénieur général des Mines and Vice-President, Bureau des Recherches Géologiques et Minières in Paris. Throughout his career he has been attached to the Service de la Carte géologique, and is a former director of the Service. He has done research and consultation on the exploitation of geothermal power since 1952,

when he served as a consultant to Electricité et Gaz
d'Algérie on the possibility of exploiting the hot springs at
Hamman Meskoutine in Algeria.

SYDNEY P. CLARK, JR.

geothermics

CHAPTER ONE
general information

   Geothermics can be defined as the study of the temperature distribution in the earth and the phenomena which influence that distribution. At the earth's surface and in the
atmosphere, complex meteorological events occur which are
strongly influenced by temperature. These phenomena will
only be mentioned briefly.

   Large areas of geology are frequently discussed without alluding to the temperature: In the study of sedimentary rocks it suffices to know that the temperature has not
been raised to the value which would allow the recrys-
tallizations characteristic of metamorphism. In hydrogeol-
ogy, permeabilities are often defined for a local temperature
without separating the role of water's viscosity (which
decreases rapidly when the temperature increases) from
the role of the rock matrix. This is in contrast to petroleum
geologists, who define the permeability of the matrix in
darcys, thus taking the viscosity of the fluid into account. A
darcy is the permeability giving rise to a flux of $1 \text{ cm}^3/\text{cm}^2 \cdot \text{s}$
in a liquid with a viscosity of 0.01 poise under the influ-
ence of a pressure gradient of 1 atm/cm. Thus, geothermics
is a relatively neglected field of geophysics, at least at moderate depths.

   On the other hand, the role of temperature is evident in
metamorphism and in volcanic phenomena; however, we are
far from reaching a satisfactory group of hypotheses to
explain these temperature variations.

   In the models which have been constructed to represent the earth's interior, although the distribution of densities and pressures is well known, the temperature distribu-

tion is extremely uncertain. Yet it is clear that the temperature distribution must play an essential role in the events which take place below the earth's surface.

In this book, we will chiefly deal with the temperature distributions found at moderate depths and the ways these distributions can be modified by circulation and changes of state of water. We will indicate methods of utilizing heat stored in rocks and of economically extracting geothermal energy, mainly through vaporization of groundwater.

For the internal regions of the earth, we will limit ourselves to the main features, without discussing in detail the several hypotheses which have been formulated.

1-1 Review of some laws of physics

One body is said to be warmer, or to have a higher temperature, than another if it spontaneously gives off heat to the other body. Definite scales of temperature have been established with instruments (mercury, and later, gas thermometers) through the use of two easily reproducible fixed points, 0 and 100°C. The Fahrenheit temperature, in °F, is defined in terms of the Celsius (formerly called centigrade) temperature, in °C, by the relation °F = 32 + 1.8°C.

By means of the second law of thermodynamics, which we will discuss later, an absolute scale of temperature can be defined. It can be shown that there exists a lower temperature limit which cannot be transgressed and which corresponds to − 273.15°C on the gas-thermometer scale. It is often convenient to measure the temperature from absolute zero while conserving our original definition of the degree (100°C difference between melting ice and boiling water at atmospheric pressure). Thus, there is a constant difference of 273°C between the absolute temperature which is measured in degrees Kelvin (°K) and the Celsius temperature. This should be carefully noted, as confusion may occur when reading certain tables.

We cannot be interested in temperature without also considering the transfer of energy which causes temperature to vary. This energy can be in the form of heat; the degradation of all other forms of energy (mechanical, electrical, electromagnetic, chemical, etc.) also produces heat. However, the first law of thermodynamics, the equivalence principle, indicates that there is a total conservation of energy with the equivalence relation that $1\,cal = 4.18\,J$. Thus, it would seem logical to measure heat quantities in joules like all other forms of energy. Nevertheless we will not do this due to convenience and present usage in the field. Recall that $1\,cal$ is the quantity of heat necessary to raise $1\,g$ of water $1°C$ at around $15°C$. The kilocalorie, kcal, is also used and is a thousand times larger than the calorie.

For a substance, the specific heat is the quantity of heat necessary to raise $1\,g$ (or $1\,kg$) of the substance $1°C$ in temperature. This specific heat is easy to measure by first raising a body to a known temperature and then placing it in a water calorimeter. The specific heat depends on the temperature; in general, for rocks, the specific heat increases with temperature and can double its room-temperature value at 500 or 700°C. The values indicated by tables[1] vary only slightly for different rocks; granite is 0.155, limestone 0.16 to 0.23, and sandstone 0.19 to 0.22. This number must be multiplied by the density to obtain the heat capacity per cubic centimeter (C), which is thus of the order of 0.4 to 0.5.

1-2 Heat transfer

Heat transfer can take place essentially through three processes: *conduction*, in which the transfer occurs gradually through a body that may be solid; *convection*, in which a fluid moves in a closed circuit and carries heat. This movement may be caused by differences in fluid density which result from differences in temperature, and thus convection transports heat upward and not downward. The final type of heat transfer is by *electromagnetic radiation*.

[1] The best source is in Sydney P. Clark (ed.), "Handbook of Physical Constants," Mem. 97, Soc. Geol. Am., 1966.

1-3 Conductive transfer

In a standing body which may be solid, if the temperature is not uniform there will be a heat transfer which we can characterize as a flux. Flux, in this case, is the quantity of heat which passes through an imaginary surface per unit area per unit time. It can be shown that this flux is a vector, which means that our definition corresponds to the flux of a vector, whatever the orientation of the surface element. The flux would be zero if the temperature were uniform. If the temperature is not, then the flux is linearly related to the temperature gradient and we may write $\mathbf{q} = K \overrightarrow{\mathrm{grad}}\, \theta$, where $\overrightarrow{\mathrm{grad}}\, \theta$ is the vector with components $\partial\theta/\partial x$, $\partial\theta/\partial y$, $\partial\theta/\partial z$.

In a crystal, and in general for all anisotropic bodies, the thermal conductivity K is a tensor. Thus, usually the flux and temperature gradient vectors do not have the same direction (except along the three principal axes, which are mutually orthogonal). The anisotropy of this tensor can be considerable: in mica, the conductivity can be as much as five times larger parallel to the cleavage plane than at right angles to it. For schists, this difference can be a factor of 2.

Even for an isotropic rock for which the conductivity K is a scalar, the measurement of the conductivity coefficient requires care, and the result will only be precise within a few percent. The most accurate method appears to be that called the *divided-bar* method, the apparatus for which is made up of two metallic supports with the same cross-sectional area and a rock cut in the form of a disk (see Fig. 1-1). The upper part of this device is heated while the base is cooled, and thus a steady-state temperature distribution is established. Then the temperature measurements are taken at several points along the metallic bars, thus giving the thermal profile. The experiment is performed again with two different thicknesses of the same rock to eliminate the effect of the temperature discontinuity which could exist at the contacts. The results are compared with quartz disks cut

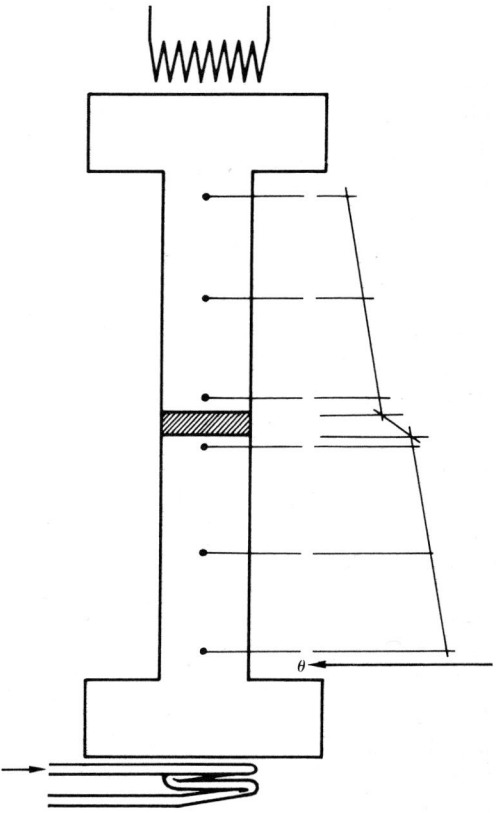

FIGURE 1-1 Principle of the divided-bar method for the measurement of thermal conductivity. The sample is cut in the form of a disk having the same cross section as the metallic rods. The rods are cooled at their bases and heated at their tops and steady-state temperatures are established. The temperatures are measured at three points by thermistors in each of the rods. The graph indicates the schematic form of the thermal profile which serves to compare the thermal gradients in the sample and the metallic rods for the same flux.

perpendicularly to the optic axis, whose conductivity is known and is conventionally taken as a reference.

The following values are cited: granite 5.7×10^{-3} cal/cm $\cdot$ s $\cdot$ °C, dunite 12×10^{-3}, limestone 5 to 7×10^{-3}, chalk

2.2×10^{-3}, marble 7.5×10^{-3}, dolomite 10×10^{-3}, quartzite 13×10^{-3}, schist 2 to 4×10^{-3}, and rock salt 27×10^{-3}. These orders of magnitude are all quite similar and much smaller than the conduction for metals (iron 0.15, brass 0.93, silver 1.001). Other less precise methods use variable temperatures. One of these will be described later.

The values for the heat capacity and thermal conductivity for rocks vary so little that, in the absence of more precise data, we can make calculations which will give an indication of the behavior of observed phenomena by taking average values. In the examples, unless otherwise indicated, $C = 0.5$, $K = 0.005$, and thus $\alpha = K/C = 0.01$.

1-4 The heat equation

In the absence of a heat source, it is easy to establish the thermal balance of an elementary volume. This balance will consist of the heat absorbed by the warming of the body and the excess of heat gained over the heat lost by conduction, $C\ \partial\theta/\partial t = -\text{div} \cdot \mathbf{q}$; for an isotropic body, $q = K\ \overrightarrow{\text{grad}}\ \theta$ which yields the equation

$$C\frac{\partial\theta}{\partial t} = K\ \overrightarrow{\text{div}} \cdot \overrightarrow{\text{grad}}\ \theta$$

$$= K\left(\frac{\partial^2\theta}{\partial x^2} + \frac{\partial^2\theta}{\partial y^2} + \frac{\partial^2\theta}{\partial z^2}\right) = K\nabla^2\theta \tag{1-1}$$

The ratio $\alpha = K/C$ is often called the diffusivity. If heat is released in the rock because of the radioactivity of some of its constituent elements, then we would need to add this heat per unit volume per unit time to the right-hand side of Eq. (1-1).

To solve a conduction problem for a solid body, we need to find θ as a function of x, y, z, and t, knowing its value at $t = 0$ and the boundary conditions, which can be either a temperature or an imposed flux.

1-5 An example of a solution of the heat equation

The principle of measuring the thermal conductivity by a needle probe demonstrates an application of the heat equation. In an indefinite, isotropic, homogenous mass initially ($t = 0$) at a constant temperature (that we may take as zero), a quantity of heat is produced at a constant rate of m cal/cm · s along a straight line.

The measured temperature depends only on the radial distance r, and the flux is radial because of symmetry. In cylindrical coordinates, the heat equation becomes

$$C \frac{\partial \theta}{\partial t} = K \left(\frac{1}{r} \frac{\partial \theta}{\partial r} + \frac{\partial^2 \theta}{\partial r^2} \right) \tag{1-2}$$

and $\theta = e^{-r^2/4\alpha t}/4\alpha t$ is a solution to Eq. (1-2).

The value of this solution for $t = 0$ is zero everywhere except at the origin where it becomes infinite. But, since its integral extended over an infinite surface is $\int_0^\infty \theta \cdot 2\pi r \, dr = \pi$, a constant, we can consider that the solution corresponds to the appearance at $t = 0$ of a quantity of heat πC per unit length of axis. If there has been a continuous production of heat between $-t$ and $t = 0$, the temperature distribution resulting from the heat diffusion at successive moments is given by

$$\theta = \frac{m}{\pi C} \int_0^t \frac{e^{-r^2/4\alpha t}}{4\alpha t} \, dt \tag{1-3}$$

This formula gives us the temperature evolution of a point situated at a distance r from the axis. If we make a change of variable such that $y = r^2/4\alpha t$, we obtain from Eq. (1-3)

$$\theta = \frac{m}{4\pi K} \int_y^\infty \frac{e^{-y}}{y} \, dy \tag{1-4}$$

This is a classical transcendental function called the *exponential integral* for which tables exist. It suffices for us to know that as soon as t is sufficiently large, its value becomes very close to ln γy, where $\gamma = 1.781$.

If the temperature has been measured at a distance r from the axis, a graph of θ versus ln t can be constructed. The curve rises progressively until it approaches an asymptote whose slope is given by $d\theta/d$ ln $t = m/4\pi K$. Or if we convert to log (to the base 10) this is 0.188 m/K. Thus we obtain the value of K by measuring the slope of the asymptote.

Conduction measurements following this principle have been chiefly done in marine sediments. A hollow needle, like a hypodermic, is equipped with an electric resistance wire passing through its axis giving off a known power dissipation of m cal/cm · s (see Fig. 1-2). The temperature is measured with a thermistor placed in a small cavity on the surface of the needle. The needle probe is then pushed into the sediment, and an electric current flows which gives off a constant thermal power; temperature is registered by the thermistor. A few minutes suffice for the measurement. If the dimensions of the apparatus are increased, the measurement time grows as the square of the diameter and the total power dissipation grows at least as the cube of the diameter, which in practice excludes the use of drilled holes with diameters of a few centimeters.

1-6 Convective transfer

Convection generally means the transport of heat by means of a free movement of fluid between two surfaces at different temperatures. Convective power is supplied by density differences between warm and cold fluid, and since the former tends to rise, convection only transports heat from the bottom to the top of the thermal system. In addition, it is not sufficient that the temperature be only slightly higher at the base of the fluid than at the top in order that convection may take place. The fluid, especially if it is a gas, cools by expansion as it rises, and thus it is necessary that the thermal gradient be at least equal to the adiabatic gradient. The adiabatic gradient corresponds to the temperature variation experienced by a mass of fluid being vertically displaced without any heat exchange. In the atmo-

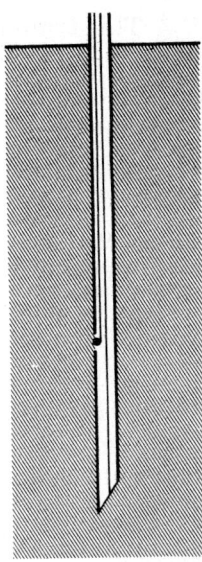

FIGURE 1-2   Principle of the measurement of thermal conduction by radial heat diffusion from one axis. A hollow needle (like a
hypodermic) is used. An axial resistance dissipates a known
constant thermal power. The temperature is measured at the
surface of the needle probe, which is sunk in the sediment to
be studied, by means of a thermistor lodged in a small cavity.
The slope of the plot of temperature versus $\ln t$ permits the
calculation of α, which gives K if C is known.

sphere, convection is the primary heat-transport mechanism, and it tends to produce a thermal adiabatic gradient
such that the atmosphere is said to be in *adiabatic equilibrium*. Two cases can be distinguished according to whether or
not the atmosphere is saturated with water.

Convection implies the individualization of ascending
currents on one side and descending currents on the other,
which can constitute more or less stable cells.

We shall frequently be concerned with a special form of
convection in which water circulates in a porous rock and is
so intimately associated with the rock that we can assume
that the water is at the same temperature as the rock at all
points. However, if there is a temperature gradient in the direction of the current lines, there is a heat transfer by water
in addition to the transfer by conduction.

This transfer is in the direction of displacement of the water, which, strictly speaking, does not correspond to the direction of its velocity because the water molecules follow sinuous paths around the grains of the rock. Nonetheless, we can characterize this flow by a *unitary-flow* vector, **u**, which is such that its flux through a surface element is equal to the mass of water which crosses the surface element per unit time. This heat transport by water is proportional to the temperature gradient in the direction of the water displacement and the heat capacity C_w of the water. Thus, the water movement results in a heat loss per unit volume of $-C_w(\mathbf{u} \cdot \overrightarrow{\mathrm{grad}}\,\theta)$, which is proportional to the scalar product of the unitary flow of the water times the thermal gradient. In practice, C_w may be taken as unity. The above term figures in the left-hand side of the heat equation in the same manner as radioactive heat production r, and so the complete heat equation becomes

$$K \nabla^2\theta + r - C_w(\mathbf{u} \cdot \overrightarrow{\mathrm{grad}}\ \theta) = C\ \frac{\partial\theta}{\partial t} \qquad (1\text{-}5)$$

Whatever the motive power for water circulation in a permeable region, this type of heat transfer can play a role. Strictly speaking, however, we can only call a heat transfer *convection* if the motive power is furnished by density variations in the water as a function of temperature. These variations vanish at 4°C, and then, of course, convection is impossible.

1-7 Radiative transfer

The only method of heat transfer in a vacuum is through radiative transfer. All bodies emit electromagnetic radiation, which carries energy, and absorb all or part of the radiation received from other bodies.

The blackbody (or the radiation which escapes from the interior of a cavity) emits radiation which depends only on the body's temperature. The energy emitted per square centimeter per second is distributed in all directions and is given by Stephan's law:

$$W = \sigma t^4 \tag{1-6}$$

where $\sigma = 5.673 \times 10^{-12}$ W/cm$^2 \cdot$ °K^4 and T is measured in degrees Kelvin.

To be more precise, the distribution of energy in the spectrum as a function of the wavelength λ, where λ is in micrometers, is given by

$$dw = d\lambda \times K_\lambda T^5 \tag{1-7}$$

with $K_\lambda = C_1/[(\lambda T)^5(e^{C_2/\lambda T} - 1)]$, where $C_1 = 1.2184 \times 10^5$ and $C_2 = 14{,}385$ when the flux is expressed in ergs per second per square centimeter. Observe that K_λ is a maximum when $\lambda T = 2897$ μ°K. The blackbody which radiates according to this law absorbs all incident radiation.

If a body is not black, it possesses for each wavelength a *color factor* which is less than 1. Its emission for each wavelength is multiplied by this factor. The energy which is received at this same wavelength is only partly absorbed, also depending on this same factor, and the rest is reflected. The color factor can vary greatly with the wavelength.

For the surface temperature of the sun (about 6000°K, the maximum for K_λ is in the visible part of the spectrum. For ambient temperatures (about 300°K), K_λ has a maximum in the far infrared at about 10 μm. Radio astronomers have established that there is also radiation from deep space which has a maximum at $\lambda = 0.7$ mm, which corresponds to a temperature of 4°K.

In a transparent substance, radiation emitted by a point source can cross a certain distance before being absorbed, and if there is a temperature gradient, the radiation emitted in opposite directions does not balance. Thus, there is a net transfer of heat by radiation. It is difficult to estimate its importance, which depends on the proportion of radiation absorbed per unit distance (which itself varies with wavelength). It appears that for certain silicates at 2000 to 3000°K, radiative transfer can be as important as conduction. At ambient temperatures, however, radiative transport is completely negligible.

In the stars, where temperatures reach tens or hundreds of millions of degrees, the radiative term is preponderant and we say that the stars may be in a *radiative equilibrium*. However, convection may play a part in the stellar interior.

1-8 Dilatation

Recall that all bodies expand when the temperature is raised. For a crystal or an anisotropic rock, expansion can be different depending on the direction. For calcite, it is negative in the direction of the optic axis.

1-9 Brief explanation of the thermal regime outside the earth's surface

Outside the atmosphere, the earth receives 2 cal/cm² · min in radiative energy from the sun. Part of this energy is reflected and part is absorbed at different heights in the atmosphere or at the ground. On the other hand, the earth radiates during the day and night following a law which differs from that of a blackbody.

We will outline a calculation which should be instructive. A spherical blackbody at a uniform temperature T is under the influence of solar radiation at the same distance as the earth. It will absorb $\pi r^2/30$ cal/s which is equivalent to $(4.18/30)\pi r^2$ W. The sphere will radiate over its entire surface $(4\pi r^2 \sigma T^4)$, and on equating these two expressions we obtain the result that $T = 280°$K, which is 7°C. Such would be the temperature of a blackbody in equilibrium with the solar radiation. This example could be that of a satellite, the color of which must be carefully chosen to bring the temperature to the desired value. The similarity between this calculated temperature and the average surface temperature of the earth cannot be the effect of chance.

But things on the earth are much more complicated than the previous example: the surface temperature at a point is neither constant nor uniform because different parts of the radiation and the absorption take place at different levels in the atmosphere (for example, in the clouds). In particular, the far ultraviolet is strongly absorbed by the upper atmosphere, which is thereby raised to a high temperature. In addition, the solar radiation varies at each point because of the rotation of the earth. Finally, the atmosphere is not static but is in a constant state of mixing which dissipates an appreciable amount of energy. Overall, these movements tend to reduce temperature variations. This means that the atmosphere functions like a thermal engine and creates mechanical energy (wind, waves, and waterpower).

1-10 Review of the second law of thermodynamics—Carnot's principle

This principle holds that perpetual motion is impossible. Now, if heat borrowed from a source at a constant temperature could be transformed into mechanical energy (inversion of the equivalence principle), perpetual motion could be achieved (that is, mechanical power could be produced from an engine borrowing heat at the ambient temperature). The second law forbids such a transformation.

Thus, mechanical energy can be produced from heat with a thermal engine only if sources at different temperatures are available. The machine will take heat from the warmer source and give it up to the cooler, but a certain proportion of heat will have been transformed into mechanical energy. The upper limit of this proportion, or *thermodynamic efficiency*, is given by a formula which also constitutes the definition of thermodynamic temperatures. If a thermal engine takes the heat Q_2 at a temperature T_2 and gives up Q_1 at T_1 while transforming $Q_2 - Q_1$, into mechanical energy, then we have

$$\frac{Q_2 - Q_1}{Q_2} < \frac{T_2 - T_1}{T_2} \quad \text{or} \quad \frac{Q_1}{T_1} > \frac{Q_2}{T_2} \tag{1-8}$$

The entropy is often defined by the relation $S = Q/T$ and we can see that the entropy gained by the cold source (Q_1/T_1) exceeds the entropy lost by the warm source. Thus, the entropy of the system can only increase. The entropy would remain constant for a system (called *reversible*) which remained infinitely close to its equilibrium point and could thereby function in either direction. Unfortunately, this is a practical impossibility. The inverse of a thermal engine would be a heat pump which absorbs heat at a low temperature and supplies heat at a higher temperature by means of consumption of mechanical power. But this time, if the efficiency is less than the thermodynamic efficiency of a reversible engine, this would mean that $Q_1/T_1 < Q_2/T_2$ and the consumed mechanical energy would be $Q_2 - Q_1$, which is larger than its ideal value.

1-11 Periodic and secular variations of the temperature

We will not deal with atmospheric and oceanic phenomena in detail but will simply recall that they influence the temperature at the earth's surface and give rise to periodic variations of which the periods of the day and the year play an important role. These variations are not sinusoidal, but we know that a periodic function (which is not exactly the case here) which is nonsinusoidal can be represented by a Fourier series, that is, a sum of sinusoidal functions with the fundamental period and the harmonics of that period.

Fluctuations with a period longer than a year are confirmed by the studies of paleoclimatology as well as the evidence of ancient glaciations. To explain these variations, should we consider the variations in the solar radiation which the earth receives? From solar studies, we are well aware of the solar cycle of 11 yr, but it does not seem to appreciably alter the radiation reaching the earth.

As a consequence of planetary perturbations, celestial mechanics allows calculation of some variations in the inclination and other elements of the terrestrial orbit. Milankovich studied the influence of these effects on the solar radiation and demonstrated that they would only lead to an average temperature change of at most 0.5 to 1.0°C, depending on the latitude.

It is also necessary to consider the possibility of instabilities (stochastic processes) in the differences of climate from year to year. The climate of one year can be influenced by the preceding year, which can lead to successive years with a warmer or colder climate than the average.

Consider a model of a globe in equilibrium with the sun's radiation. We will give an example of a thermal instability. Surfaces covered with snow or ice reflect a considerable portion of the solar radiation centered in the visible part of the spectrum. On the other hand, in the far infrared (10 μm), ice and snow are very absorbent and radiate like a blackbody. This is the wavelength of maximum thermal emission at about 0°C. Thus, if the snow- and ice-covered regions increased, the proportions of solar energy received would diminish but not the energy radiated by the earth. Therefore, the average temperature would decrease, leading to more surface area being covered with ice and snow, etc. This effect, of course, is much more complex, and such a rudimentary analysis cannot predict the rate at which the temperature would vary and the amount of surface which would be covered with ice and snow. It is also possible that opposing phenomena exist, caused, for example, by atmospheric movements, which could limit such instabilities. The complexity of phenomena such as those listed defies, even today, a precise analysis.

We can mention other examples of instabilities. Suppose that the accumulation of ice raises the land surface, until it reaches an altitude where the temperature is so low that an ice cap is preserved. (Compare Greenland, with an ice cap of about 3,000 m in elevation, with Baffin Island or other islands of the northern Canadian Archipelago.) Furthermore, such a large ice cap tends to create anticyclones which further lower the temperature. We will not enter into detail on such questions since they are poorly understood. Simply remember, we should not treat the average annual temperature as a constant since it can easily undergo short-

or long-term fluctuations. In addition, there are effects of man-made thermal pollution, which raise winter temperatures several degrees in cities, cause the heating of rivers, and lead to the formation of smog which perturbs solar radiation.

In the geologically recent past, such fluctuations resulted in glacial periods; for the last glacial period, the Würm, enormous glaciers covered Scandinavia, parts of Northern Europe, the Alps, the larger part of Canada, and the northern part of the United States and did not disappear for tens of thousands of years. The temperature perturbations of the ground must have been stronger in the periglacial zones than in those areas protected by glaciers. The Quaternary glaciations seem to have begun 1 or 2 million yr ago.

Insofar as we know, there were no prior glaciations of major importance before the one just listed. However, there have been traces of some glaciations for certain geological periods: the upper Precambrian, the Ordovician (Sahara), and the Permo-Carboniferous (South America, South Africa, Australia).

1-12 Temperature of the ground

Up to this point, we have mentioned the temperature determined by atmospheric phenomena and solar radiation at the earth's surface. These influences produce temperature fluctuations of short periods around an average value which also fluctuates. If the temperature of the ground is measured at only a few centimeters depth, one finds that even in a limited area the temperature is strongly dependent on local conditions: Bare ground warms up more in the sun than land which is protected with even herbaceous vegetation. Such vegetation can evaporate water from the soil, which has a cooling effect, and, depending on the nature of the vegetation, will, or will not, affect the ground. The ground can be either favorable or not to dew deposits or hoar frost accumulations. These phenomena defy analysis, and we are obliged to consider the temperature of the ground at a few centimeters in depth as empirical data.

1-13 Propagation of periodic temperature variations into the ground

Let us imagine a homogenous earth with an infinite plane surface. By reason of symmetry, the temperature depends only on the depth, and thus the heat flux is vertical. It is sufficient to write the heat equation for the vertical co-ordinate only:

$$\alpha \frac{\partial^2 \theta}{\partial z^2} = \frac{\partial \theta}{\partial t} \qquad (1\text{-}9)$$

The boundary condition will be the temperature at the earth's surface for $z = 0$. If this temperature is expressed by a sum of periodic terms, we suppose that the expression of temperature will be of the same form at some depth in the ground after a sufficient time. This condition replaces the specification of an initial temperature.

It is immediately seen that all solutions of the form $\theta = A \exp(-z\sqrt{\omega/2\alpha}) \sin(\omega t - z\sqrt{\omega/2\alpha})$ satisfy Eq. (1-9). There would also be a solution with terms of the form $\exp(+z\sqrt{\omega/2\alpha}) \sin(\omega t + z\sqrt{\omega/2\alpha})$, but it would be absurd to find temperature variations increasing in amplitude with depth. If the surface temperature is represented by a sum of periodic terms, each of these terms will correspond to a term in the solution.

It is seen that temperature variations of a sinusoidal nature propagate downward while decreasing in amplitude. For $\alpha = 0.01$, an average value for rocks, the amplitude is reduced in the ratio by $e^{-\pi} = \frac{1}{23}$ at a depth $\sqrt{\alpha \pi T}$, T being the period in seconds. For $T = 1$ day, a depth of 52 cm is found, and for 1 yr, the depth is $\sqrt{365}$ times larger, or 10 m. At the same depth, the wave lags by π, i.e., is in opposition to the surface variation, since the propagation entails a delay of a half period. At twice the stated depth, the amplitude would be reduced in the ratio $(\frac{1}{23})^2 = \frac{1}{529}$. This is completely negligible (see Fig. 1-3).

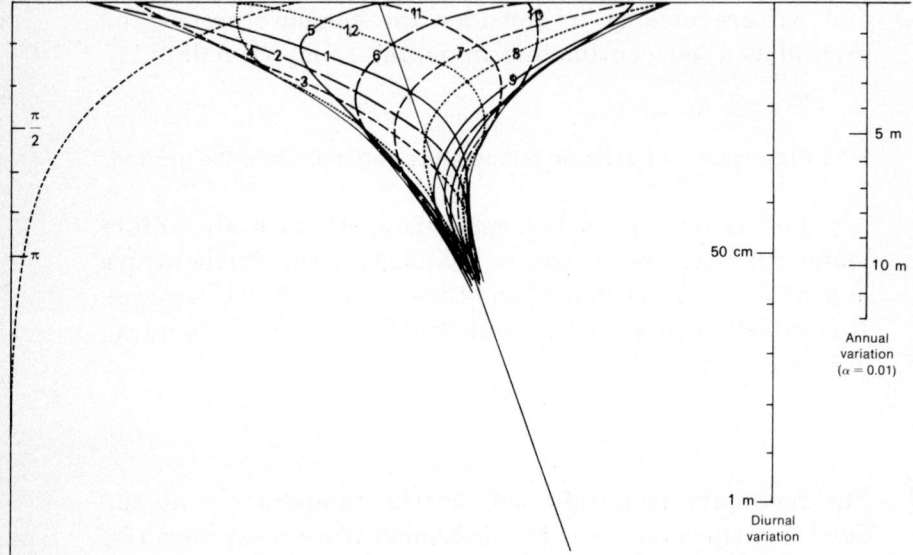

FIGURE 1-3   Temperature profile in a homogenous soil for a sinusoidal variation of the surface temperature, including the subterranean heat flux. The 12 curves can represent the monthly profiles for the annual variation, or every 2 hours for the diurnal variation. The vertical scales are marked for $\alpha = 0.01$ cm^2/s.

These conclusions are amply confirmed by experiment. We may add that if the surface variations are not sinusoidal, the higher harmonics decrease more rapidly and the temperature variation underground tends to become sinusoidal.

1-14 Stratified ground

If the ground were composed of horizontal layers with different thermal properties, there would be no difficulty in calculating the propagation of periodic temperature variations. For each layer, solutions of the form indicated for Eq. (1-9) would be valid, without excluding the solutions $\exp(+z\sqrt{\omega/2\alpha}) \sin(\omega t + z\sqrt{\omega/2\alpha})$. These solutions can be considered as reflected waves at the interfaces which propagate upward. For each interface, the temperature and flux equalities can be written for the two separate layers; each equality includes the phase and the amplitude for a sinusoidal term with a given period. The phase and amplitude for the two types of solution can be successively determined at each interface. The results of these calculations can be expressed by saying that at the interface crossing, waves traveling in the same direction have their amplitude multiplied in passing from layer 1 to layer 2 by $\frac{1}{2}(1 + \sqrt{K_1 C_1 / K_2 C_2})$ and that at the same time, a wave is reflected in the opposite direction with an amplitude multiplied by $\frac{1}{2}(1 - \sqrt{K_1 C_1 / K_2 C_2})$. Begin with an arbitrary phase and amplitude from the deepest, indefinite layer in which there exists only one type of solution. Afterward, determine the solution in each layer until the surface is reached, where the phase and amplitude are adjusted to meet the imposed boundary conditions. The general behavior of the solution stays the same as for a homogenous region.

Secular variations in temperature, the details of which are poorly understood, could in principle be described by a Fourier series, each term of which would easily allow us to calculate the underground effects. The depth penetration, that is, the depth at which the amplitude is reduced in a specified ratio, is proportional to the square root of the period. Considering the distance $l = \sqrt{2\alpha/\omega}$, which is of the order of $3\,\text{m} \times \sqrt{\text{period in years}}$ and for which the amplitude is reduced by the ratio $e = 2.7$, it is interesting to observe that the local gradient, which naturally decreases with depth as the temperature variation, has as a surface maximum one-half the amplitude of the total variation divided by $0.7l$.

As an example, for a period of 100,000 yr, $l = 1,000$ m. If the amplitude of temperature variation was 10°C, the maximum surface gradient would be 0.007°C/m and at 1,000 m, 0.003°C/m. The total amplitude of the temperature variation at the latter depth would still be 3.7°C. In a geothermal flux measurement down to 1,000 m, the error introduced could be of the order of 10 percent.

The variations can be more rapid and irregular, thus introducing terms with shorter periods. This effect will be eliminated if the lower parts of a drill hole are used, which is also recommended for other reasons.

1-15 Stresses caused by temperature variations

If a solid body was free from mechanical stresses, it would dilate as the temperature changed. When external conditions impose a certain temperature distribution on a body (we shall see how to calculate this), the geometric and mechanical solidarity of different parts of the body do not allow it to dilate freely. A distribution of mechanical stresses can result from this, which it is interesting to calculate. Of course, these stresses will be added to the preexisting stresses which we will not take into account here.

These internal stresses of the body are such as to be in equilibrium with the zero external forces. They produce on each volume element an elastic deformation, given by the classical equations, which adds to the thermal dilatation. It is the sum of these two deformations which make up the real deformation, subject to the geometric conditions, that they derive from a displacement. If we know the temperature distribution, the mathematical theory of elasticity will allow us to calculate the distribution of the thermal stresses and the deformation.

Actually, this procedure is not rigorous. Elastic compression releases heat, and if it takes place quickly, the temperature will rise. It would be necessary in the case of a rapid deformation (in particular, in the study of sound waves and especially shock waves) to write the mechanical and heat equations together.

Nonetheless, we will be content to take the approximation corresponding to slow changes, where we can by some means independently determine first the temperature distribution and then the resulting mechanical stresses.

As a case in point, let us consider the following problem. A blackbody (such as a satellite) is in thermal equilibrium with the solar radiation. An eclipse cuts off this radiation. How will the temperature change near the satellite surface which is turned toward the sun? (We suppose it is not rotating.) What mechanical stresses will result?

We will take $C = 0.5$, $K = 0.005$, and thus $\alpha = 0.01$, the average value for rocks. Now, as a first approximation, we will consider that the radiative flux loss is constant and equal to $\frac{1}{30}$ cal/cm² · s. We will let the temperature T (°K) be represented by $T_0 - \theta$; thus, $\partial\theta/\partial t = \alpha\,\partial^2\theta/\partial x^2$, with $\theta = 0$; at $t = 0$ and at $x = 0$ an imposed flux $K\,\partial\theta/\partial x = \frac{1}{30}$.

Considerations of homogeneity lead us to try a solution of the type $\theta = A\sqrt{\alpha t}\,\Phi(z)$, with $z = x/\sqrt{\alpha t}$. The equation then becomes $2\Phi'' + z\Phi' - \Phi = 0$. Let $\Phi = fz$ and we obtain $f''/f' = -(4 + z^2)/2z$ and thus $\ln f' = -\ln z^2 - z^2$. Integrating, we obtain $f = e^{-z^2}/z - \sqrt{\pi}\,[1 - \mathrm{erf}\,(z)]$, where $\mathrm{erf}\,(z) = (2\sqrt{\pi})\int_0^z e^{-y^2}\,dy$ is a function which often appears and which is tabulated in the appendix. Thus, $\Phi = fz = e^{-z^2} - z\sqrt{\pi}\,[1 - \mathrm{erf}(z)]$ and $\Phi' = -\sqrt{\pi}\,[1 - \mathrm{erf}\,(z)]$. For $x = 0$, $\partial\theta/\partial x = -A\,\sqrt{\pi}$, from which we obtain $A = 3.78$.

The surface temperature is given by $\theta = 3.78\sqrt{\alpha t}$, which is 3.78°C after 100 s, 37.8°C at 2 h 46 min, etc. However, after several hours we must take the slowing down of the cooling rate by radiation into account.

In the calculation of mechanical stresses, the pressure on the free surface is, and remains, zero. But a tension parallel to the surface appears which corresponds to an elastic elongation, equal and opposite to the free thermal dilatation.

We shall take for the expansion coefficient the mean value for quartz, that is, $\delta = 0.13 \times 10^{-4}$, and elastic coefficients $\lambda = \mu = 25 \times 10^{10}$ dyn/cm². Now we need to write an equation containing the three elasticity equations: $\sigma_{ii} = \lambda(\epsilon_{11} + \epsilon_{22} + \epsilon_{33}) + 2\mu\epsilon_{ii}$. Letting $\sigma_{ii} = 0$, $\sigma_{22} = \sigma_{33}$, and $\epsilon_{22} = \epsilon_{33} = -\delta\theta$, we solve the equations and obtain the result $\sigma_{22} = \sigma_{33} = (4 - \frac{2}{3})\lambda\delta\theta$, which is a stress of 11.44 bars at the end of 100 s, 114.4 bars at 2 h 46 min, etc. It remains to be seen whether these tensions, which can be repeated a number of times, can cause a surface fragmentation of the body.

heat flow

All the considerations of Chap. 1 could apply to an earth with a thermally inert interior. Such is not the case, although the perturbation of surface temperature by subterranean heat is negligible. This is easily demonstrated by considering the schematic model of a spherical blackbody which is in thermal equilibrium with the sun's radiation. We found, in Sec. 1-9, that such a body reaches a temperature of $280°K$. If we modify this model by supposing an internal heat flux equal to the average geothermal flux (i.e., $1.2 \ \mu cal/cm^2 \cdot s$), Stephan's law immediately yields $\Delta W/W = 4\Delta T/T$; the mean power dissipation for the whole surface of the earth being $\frac{1}{4} \times 30 \ cal/cm^2 \cdot s$, it is increased by heat flow by 1.44×10^{-4}, which gives a temperature variation $\Delta T = 280°K \times (1.44 \times 10^{-4}/4) = 0.01°K$.

Even though this blackbody model is crude in comparison to the complexity of the external thermal system, we should remember that internal thermal activity cannot raise the surface temperature of the earth more than about a hundredth of a degree. This, of course, disregards unusual points such as next to volcanoes or active thermal sources.

This comparison emphasizes the fact that surface energy exchanges, where mobile fluids interact, are infinitely more intense than those in the earth's interior.

2-1 The measurement of heat flow

It has long been realized that once a depth of 10 m is exceeded (the effective limit of annual temperature perturbations), there is a steady increase in temperature on the order of 1°C for each 30 or 40 m. Sometimes, to designate this distance, the improper expression "geothermal degree" or "geothermal step" is used. This temperature increase with depth had been observed in mines in the seventeenth century and was attested to by Boyle in 1671 and Gensanne in 1740, although the temperature measurements were difficult to perform then. Arago was interested in this phenomenon in 1852, but it was not until 1868, under the influence of Lord Kelvin (then William Thomson), that a committee of the British Association systematically collected data on the thermal gradient and the thermal conductivity values for various rocks. Although these two types of measurements were not carried out in the same places, it was deduced that the average heat flux transmitted by conduction to the earth's surface was 1.3 μcal/cm^2 · s, an order of magnitude which has subsequently been confirmed.

In 1935, a new committee under the same association had Benfield and Bullard perform the measurements. Thermal profiles were measured in boreholes, and samples from the holes were used to determine the thermal conductivity. From that time on, it has been usual to present a graph with the measured temperature as the abscissa and the *thermal resistance*, $\int dz/K$, as the ordinate. Such a plot usually produces a straight line. It thus appears that the heat flow is approximately constant in a mine shaft or boring, even if the shaft crosses quite different rock layers. The geothermal degree (or gradient) is much more variable.

It was also demonstrated that for the first 1,000 m (in which most of the measurements were performed), a correction factor to take account of the earth's cooling during glacial periods would be on the order of 10 percent. This number is impossible to calculate accurately though, because we would need to know the detailed thermal history of the earth's surface. This effect tends to make us underestimate the heat flow.

Around 1949, Bullard made the first geothermal flux measurements in the ocean. It was expected that the heat flow there would be much lower than that existing on the continents, but this was not the case and a complete new view of models explaining the relationship between the oceans and the continents had to be taken. Taking advantage of the fact that the bottom of the ocean is generally at a constant temperature near 0°C, probes were dropped into the ocean bed. These probes were 2 or 3 m in length and had electrical devices at three points to measure the temperature. In the head of the probe was a recorder, and the measurements were delayed for the longest time possible (at a minimum, $\frac{1}{2}$ hour). During the recording of the temperature by the three thermometers, first the temperature of the seawater was seen (which allowed the thermometers' mutual calibration to be checked), then a warming due to friction as the probe penetrated the sea floor, and finally a restoration period during which thermal equilibrium was approached at an exponential rate. It was necessary to allow for a long enough time to elapse to extrapolate the rate of this curve and determine the equilibrium temperature of the oceanic sediment. If a sample of the ocean bed was obtained, then the thermal conductivity could be determined with a needle probe or deduced from the water content of the sediment by an empirical formula.

Ewing improved Bullard's apparatus by placing the temperature sensors on lateral fins fixed to the probe, which greatly accelerated the return to thermal equilibrium.

Admittedly, it is difficult to know if the probe is planted vertically in the sediment if the probe is not equipped with an inclinometer, and this can lead to an underestimate of the flux if the probe is inclined.

Thousands of measurements have been taken in the ocean with a precision which does not exceed 10 percent. These measurements have provided the same average heat-flow value as for the continents (1.2 μcal/cm$^2 \cdot$ s), but with significant regional variations to which we will later return.

Further continental heat-flow measurements were performed later; under the auspices of the International Union of Geodesy and Geophysics in 1963, measurements were made in many countries, including France. Drill holes which originally had served other purposes were used. Obtaining rock samples of the rock strata traversed was not too difficult to achieve, but drilling operations dissipated energy and heated up the surrounding area. In addition, the circulation of mud completely modified temperature distributions. This allows use of thermal measurement to abstract some information for various conditions. For instance, the depth of permeable layers from which water can penetrate into the shaft will be clearly indicated on a profile measured some time after the temperature has been stabilized by cooling mud in the drill hole. Another application is to determine the height of a cement collar which is being placed around the casing by observing the heat released in the cement's solidification.

To obtain the thermal profile of the surrounding ground, it is necessary to leave the hole alone for a sufficient amount of time. The ideal situation would be to wait for a time equal to the actual drilling operation. Otherwise, measurements of the thermal profile can be taken at intervals of a few days, and from this the temperature evolution can be extrapolated.

The first measurements were made by lowering a maximum thermometer or, even better, a nongraduated thermometer whose shaft was cut obliquely. The mercury flowed out, and by heating the thermometer in the laboratory again, the maximum temperature could be calculated. These thermometers were placed in containers to protect them from the pressure.

Today, we prefer to do electrical measurements with a thermistor the resistance of which varies greatly with the temperature, but which must be specially calibrated. An elegant technique consists of regulating the frequency of an oscillating circuit by means of a thermistor, where the entire circuit is at the bottom of the drill hole, protected from the pressure. It is this oscillator which is standardized and the connecting cable has no influence on the measured frequency. With such a system, a precision of 0.01°C can be realized. However, this precision is perhaps illusory because it represents the temperature of the cooling mud, not the surrounding rock.

In New Zealand, for drilling operations which were at too high a temperature to allow reliable electrical measurement, an instrument was used which was inspired by the bathythermograph of oceanographers. This is a thin bimetallic strip which bends with temperature and which moves a needle which leaves a trace on a smoked-glass plate. The plate is connected to a ratchet device which advances the plate with each shock given to the cable which lowered the apparatus. The bimetallic strip and associated elements are standardized in the laboratory at the same pressure as in the drill hole.

The most notable cause of error is certainly due to aquifers that the drill hole penetrates. These layers are generally not in equilibrium with one another and give rise to a circulation along the hole. If there is a casing, such flows can circulate around it. To detect such errors, it is indispensable to measure a continuous profile or at least to take a large number of measurements. However, if such circulations are present, it is doubtful that their effect can be eliminated solely by calculation. The ideal situation would be to take the measurements in an impermeable homogenous layer. Too often though, we only have borings in sedimentary areas or near mineral deposits where there are frequently complex structures. Few authors have the courage to reject, as doubtful, measurements which required much time, money, and effort to obtain. Published heat-flow values undoubtedly include many doubtful measurements which are difficult to recognize.[1]

Nevertheless, large areas still lack any measurements and it would be completely illusory to seek overall patterns of heat-flow distribution on a worldwide scale. We will simply note that the heat flow appears low on ancient shields (approximately 1 μcal/cm$^2 \cdot$ s) and a little higher in recent

[1]See William H. K. Lee (ed.), "Terrestrial Heat Flow," 276 pages, no. 9, Geophysical Monograph Series, American Geophysics Union, 1965; also Geothermal Problems, *Tectonophysics*, vol. 10, nos. 1–3, September 1970.

tectonic zones. In the ocean, there are good indications of higher values along and near rifts on the midoceanic ridges (up to 8 or 10 μcal/cm^2 · s). On the volcanic island arcs of the Pacific, low values (about 1) are found at the outside of the arcs and high values (about 2) on the inside. Any structural model of island arcs will have to take these variations in heat flow into account.

Of course, in the regions near active volcanoes, and to a lesser degree near thermal sources, all kinds of values for the heat flow can be found, even very high ones. However, it is clear that these active volcanic regions contribute only a negligible amount to heat flow on a global scale. The heat radiated and transferred by convection over the lava lake which occupies the crater of Nyiragongo is estimated at about 1 million kW. It is easy to calculate that this power is equal to the average heat flow (1.2 μcal/cm^2 · s) over the surface of a circle with a diameter of 160 km, which is $\frac{1}{25,000}$ the surface of the earth. It is likely there are no more than one or two volcanoes having such permanent activity of this type at this time; out of about 500 known volcanoes, the majority have only intermittent activity and are much less intense than the lava lake at Nyiragongo. For Japan, the energy dissipated annually for all volcanoes has been estimated as an average of 7.3×10^{24} ergs/yr which is 2.3×10^6 kW, twice the value for Nyiragongo. That is one-tenth the normal heat flow for all Japan, and hot ground and thermal sources bring a surface flux of about the same magnitude.[1] But, by all evidence, Japan is not a representative case. Thus, it is clear that the major portion of the heat which escapes from the earth's interior is due to the normal heat flow. The additional heat supplied by volcanic zones is much smaller, on the order of $\frac{1}{1,000}$ or $\frac{1}{10,000}$. A precise figure is very difficult to give. What was the flux like during the time of formation of large basaltic plateaus (Deccan, Siberia, South America, Ethiopia, etc.), where the surface cov-

[1]Kozo Yuhara, Heat Transfer Measurements in a Geothermal Area, *Tectonophysics*, vol. 10, pp. 19–30, 1970.

ered with lavas may have approached millions of square kilometers? We do not know the durations of such outpourings and thus an estimate of the heat loss is difficult. Observe that the cooling of 1 million km³ of basalt from 1000°C represents a heat loss on the order of 0.5×10^{24} cal. This corresponds to the normal heat flow for the whole world for a period of 2,600 yr. If the release of the basalt had lasted for a much longer time, the heat loss would still be small in comparison with the total normal heat flow.

It is easy to calculate that the total power dissipated by the normal heat flow is equivalent to 2.55×10^{10} kW. Of course this is much less than the energy brought to the earth by solar radiation (1.7×10^{14} kW), but most of the solar energy is either reflected or immediately reemitted.

We will see in Chap. 8 that the heat flow represents much more power than is dissipated by other manifestations of internal activity such as tectonic deformations or earthquakes. Also, note that the energy dissipated by tidal actions, which is taken from the kinetic energy of rotation of the earth and moon, is roughly estimated at 3×10^9 kW, which is approximately one-tenth of the geothermal flux.

2-2 Structural influences on the heat-flow distribution

We have implicitly assumed that the surroundings are homogenous in our descriptions of measurement methods, and that heat can be considered to flow vertically. This is not necessarily so, and the lack of knowledge about local conditions can lead to errors. If different thermally conducting layers are joined vertically, it is the thermal gradients which are equal while the local fluxes are different. The regional flux, significant in subterranean regions, is a weighted average of the local fluxes. On the other hand, for horizontal layers with different thermal conductivities, the flux is the same and the thermal gradients are different. For other more complicated arrangements, neither the flux nor the gradient is uniform, or even vertical. It is perhaps the method of the electrical analogs which would most easily permit us to determine the temperature distribution (analog of the potential) of a complex structure.

Assuming a homogenous, horizontal, transgressive terrain covering a complex structure with different rock units joined along vertical surfaces, the value of the gradient and the flux in the covering material furnishes an indication of the underlying rock distribution, since both thermal quantities are greater above highly conducting rocks. Theoretically, this would provide a prospecting method were it not so difficult to make precise measurements of the thermal gradient.

For a vertical flux, it is also necessary that the surface temperature be uniform. A borehole that was sunk at Resolute Bay, in the northern Canadian Archipelago, was measured for its thermal profile (in this case, the freezing of water around the measurement cable avoided the risk of perturbations caused by water circulation), and it appeared that there was an abnormally high flux. But in fact the site was next to the ocean, and if the average land-surface temperature was of the order of $-20°C$, and the temperature at the bottom of the ocean stayed constant at about $0°C$, then in this case, the flux was certainly not vertical.

To study this case's behavior, we will schematize the situation by supposing that at the surface there were two temperatures differing by $\Delta\theta$ on either side of a linear boundary. Assuming that at great depths the flux has a given constant value, determine the temperature distribution.

We will take advantage of the fact that, since the heat equation is linear, a sum of individual solutions to the equation will also solve the equation. We will first consider the situation where the temperature at the surface ($z = 0$) is $\Delta\theta/2$ for $x > 0$ and $-\Delta\theta/2$ for $x < 0$, the thermal flux canceling at great distances. It is immediately seen that the solution for which the isotherms are planes passing through the axis $x = 0$, making an angle A with the vertical such that $\theta^1/\Delta\theta = A/\pi$, satisfies the given conditions. The unit flux, perpendicular to the isothermal planes, has a value $K \Delta\theta \pi r$, where r is the distance from the axis $x = 0$.

We will also take into account another part of the solution corresponding to a uniform flux whose value F is determined with the condition that the surface temperature equals zero ($\theta_2 = 0$); isotherms are then given by $z = K\theta_2/F$.

In adding these two solutions, we will find for $\theta = \theta_1 + \theta_2$ a distribution which satisfies the equation $\theta = \Delta\theta A/\pi + Fz/K$ (see Fig. 2-1 indicating the shape of the isotherms). It is seen that up to a distance $l = K \Delta\theta/F\pi$, the ocean gives up heat to the bottom, the flux being locally inverted. It is only further from the shoreline that the flux has its normal direction, the value being $F(1 - l/x)$. On land, the flux $F(1 + l/x)$ locally (for small values of x) takes values which can be quite high.

From the preceding solution, we can immediately draw the solution for a large limited area such as a lake, where there is a temperature difference $\Delta\theta$ between the area in question and the rest of the surface (in France, the bottom of a lake can be at 4°C while the surface is at 11°C). It is sufficient to add the solutions above for the two limits; in the absence of a subterranean flux, the isotherms would be circles; in its presence, the isotherms have shapes such as those indicated in Fig. 2-2.

2-3 Temperature in a tunnel

A problem frequently studied because of the interest in estimating the initial temperatures of rock strata which are cut by a tunnel concerns a mountain with average ground temperatures decreasing at a rate in the neighborhood of 1°C/100 m. As usual the flux has its normal value at large distances; the temperature distribution is wanted. Only approximate solutions can be found, taking the shape of the mountain more or less exactly into account and painfully estimating the rock distributions, with different conductivities, often anisotropic. (The high temperatures encountered

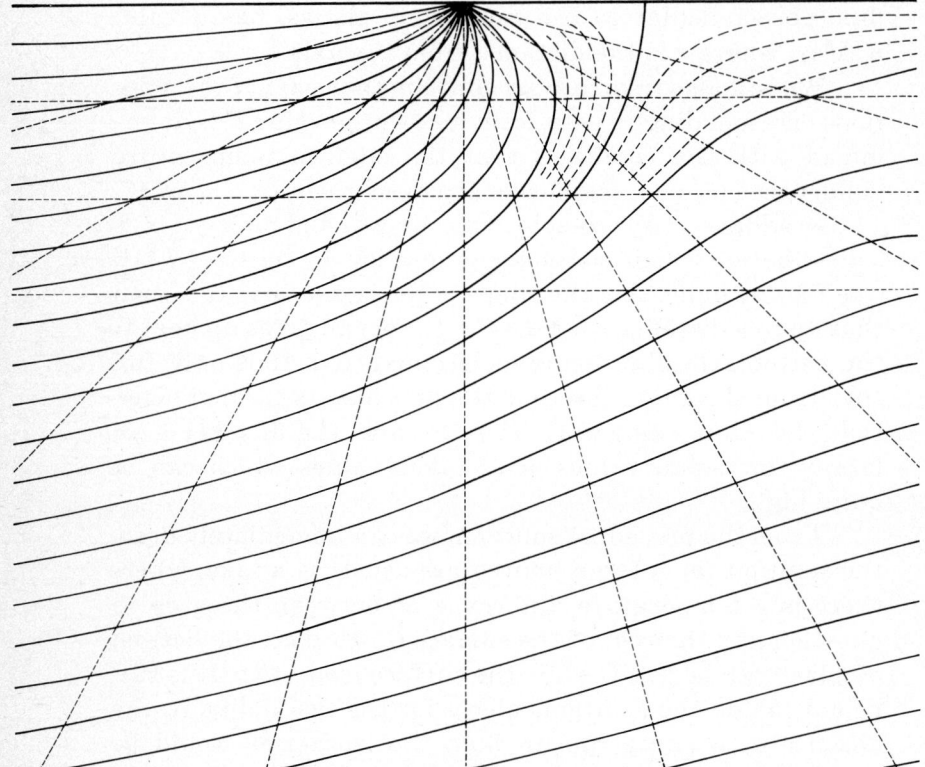

FIGURE 2-1 Temperature distribution under a rectilinear bank along
which the average surface temperature suffers a discontinuity. The graph is applicable by a scale choice to all values of
the discontinuity and the geothermal flux. (The scale can be
fixed by means of the point where the flux is zero, corresponding to an isotherm, the abscissa of which is given in the
text, which reaches the surface perpendicularly.)

in the Simplon tunnel were interpreted in retrospect as
being due to the fact that the schistosity of the gneiss was
horizontal while the Gothard tunnel's schistosity was vertical.) It can be useful to consider the desired solution as
composed of the superposition of the normal thermal gradient, such as would be the case if the earth were horizontal,
with indefinite upward extension and a correction term
which must satisfy the heat equation and the imposed sur-

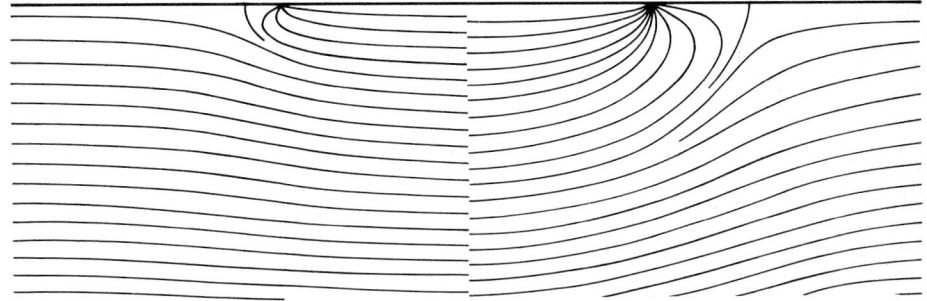

FIGURE 2-2   Temperature distribution under a lake compared to a parallel
band where the average temperature is less than that of the
surrounding surface. To the left: case where the average superficial temperature difference corresponds to the value of
the thermal gradient for a depth of 0.2 times the breadth of
the lake. To the right: for a case where the depth is equal to 0.8
times the breadth of the lake.

face temperatures, corresponding to a zero flux at infinity.
Also, the solution must be zero on a plane surface outside
the mountain and equal to the difference between the imposed surface temperature and the temperature resulting
from the extrapolation of the thermal gradient to the mountain's surface.

Among the methods which we can consider, let us pay
particular attention to electrical analogs. If we have a
mold in the form of a mountain filled with a conducting fluid
and if we establish voltage potentials along curves of equal
level, we can determine the form of the volume equipoten-
tials (or along the projected tunnel route) and thereby
deduce the correction for surrounding topography.

2-4 Effects of erosion or sedimentation

Another perturbation can result if the surface temperature remains constant but there is erosion or deposition of
sediment, which we will suppose to be of the same material
as the substratum. To study this type of problem, a moving
coordinate system is sometimes used whose origin remains

at the surface while the surface moves with a velocity u. This requires that we replace $\partial\theta/\partial t$ by $(\partial\theta/\partial t - u\,\partial\theta/\partial x)$, $u\,\partial\theta/\partial x$ being the apparent temperature increase at a depth x, caused by the movement of the origin, and not corresponding to any actual heating. The heat equation then becomes

$$C\left(\frac{\partial\theta}{\partial t} - u\,\frac{\partial\theta}{\partial x}\right) = K\,\frac{\partial^2\theta}{\partial x^2} \quad \text{or} \quad \frac{\partial\theta}{\partial t} = \alpha\,\frac{\partial^2\theta}{\partial x^2} + u\,\frac{\partial\theta}{\partial x} \qquad (2\text{-}1)$$

where u is positive in the case of erosion and negative for sedimentation, the axis being directed downward.

This equation allows us to solve the following problem: if erosion continues at a constant speed in a steady-state system, what would the temperature, which is assumed uniform, be at great depths underground in order that we observe the given surface gradient? The desired solution, which is assumed constant in a moving coordinate frame, is immediately found:

$$\theta = A(1 - e^{-ux/\alpha}) \quad \text{and} \quad \left(\frac{\partial\theta}{\partial x}\right)_0 = Au/\alpha$$

If we imagine an erosion of 1 mm/yr with $\alpha = 0.01$, the superficial gradient is 1°C/30 m which would correspond to a temperature at great depths of 950°C. However, the equivalent case for a uniform sedimentation cannot be visualized. The preceding problem is not realistic because we know that the temperature is not uniform underground and the erosion could not have followed its course indefinitely in the past (the solution of this problem would be drastically changed if the erosion had not been uniform at least for 50 or 100 million yr). At least, it gives us an idea of the degree to which erosion of long duration can disturb the thermal system. This reasoning cannot be applied if we look for a steady-state solution (in a moving coordinate system) given a condition satisfied at a great distance such as a nonzero flux or gradient, because the temperature variation of the entire half space would absorb an unlimited amount of heat flux.

A realistic way of treating the problem would be, given the steady-state solution corresponding to a given flux at a great depth, to consider erosion or sedimentation proceeding at a given rate during a limited time span and to look for the resulting perturbation of the thermal profile.

There are no analytically simple solutions in general, but if we consider erosion which goes as the square root of the time, $x_0 = p\sqrt{t}$ (p would be negative in the case of sedimentation), the solution

$$\theta = mx - A\sqrt{t}\,\Phi\left(\frac{x}{\sqrt{4\alpha t}}\right) \tag{2-2}$$

with $\Phi(z) = e^{-z^2} - z\sqrt{\pi}\,\mathrm{erfc}\,(z)$, satisfies the heat equation for fixed axes and satisfies the condition $\theta = 0$ for $x = p\sqrt{t}$ if $A = mp/\Phi(z_0)$ with $z_0 = p/\sqrt{4\alpha}$.

From these conditions, we get the value for the gradient:

$$\frac{\partial \theta}{\partial x} = m\left[1 + \sqrt{\pi}z_0\,\mathrm{erfc}\left(\frac{x}{\sqrt{4\alpha t}}\right)/\Phi(z_0)\right] \tag{2-3}$$

Observe that at the surface, the gradient is a constant. If $z_0 = p/\sqrt{4\alpha}$ is small, which will generally be the case (for an erosion of 1,000 m in 1 million yr, which is 1 m in the first year, $z_0 = 0.089$), we can use the approximation $(\partial\theta/\partial x)x_0 = m(1 + \sqrt{\pi}z_0 - z_0^2)$.

From these results, we immediately deduce that in the case where erosion progresses uniformly, the gradient will constantly increase. Also, this gradient will be greater than the gradient for the same erosion in the same time following a law going as the $\sqrt{t}$ and less than one for the same erosion in half the time following a law going as the $\sqrt{t}$ (which will finally lead to the same instantaneous velocity). This suffices to furnish an order of magnitude.

We could also do an approximate calculation by supposing that the erosion occurs instantaneously. It is easy to see that the temperature profile corresponding to a progressive erosion falls between the results of an instantaneous erosion (at the beginning) and a continuous erosion (at the end).

If the initial gradient is given by $\theta = mx$ and if at $t = 0$ a thickness l of ground is removed and the new surface is brought to the initial temperature, taken as 0°C, the temperature is then given by

$$\theta = mx + ml \ \text{erf}\left(\frac{x}{\sqrt{4\alpha t}}\right) \tag{2-4}$$

The surface gradient is $m(1 + l/\sqrt{\pi\alpha t})$. This formula shows that 1 yr after the removal of 1 mm of ground the gradient is only raised by 1 part in 10,000. But 1 million yr after the erosion of 1 km, the gradient is increased by 10 percent (these values are lower but in approximately the same ratio as would be produced by continuous erosion in the same time span).

Sometimes in place of these erosion or sedimentation problems, a more complicated problem is considered, in which we are given the temperature distribution as a function of depth (for example, $\theta = ax + T_0 (1 - e^{-bx})$, and the grounds' contents of radioactive elements decreasing exponentially as a function of depth, and we must find a steady-state solution as a function of the rate of erosion.[1] But it becomes very difficult to distinguish between the desired physical phenomena and the consequences of particular choices of the parameters T_0, a, and b, which characterize the given temperature profile.

2-5 Effect of radioactivity

Until now, we have ignored the influence of the radioactivity of rocks, which results in a heat production which can reach 20×10^{-6} cal/g · yr for certain granites (see Chap. 8). Generally it is much smaller for sedimentary rocks, of the order of 2×10^{-6} cal/g · yr.

Considering the complete heat equation,

$$C \ \frac{\partial\theta}{\partial t} = K \left(\frac{\partial^2\theta}{\partial x^2} + \frac{\partial^2\theta}{\partial y^2} + \frac{\partial^2\theta}{\partial z^2}\right) + r \tag{2-5}$$

[1]H.S. Carslaw and J.C. Jaeger, "Conduction of Heat in Solids," p. 388, Oxford University Press, 1959.

we can look for solutions of the form $\theta = \theta_1 + \theta_2$, where θ_1 is the solution to Eq. (2-5) without the term r, and θ_2 is a solution to the equation $0 = K\nabla^2\theta + r$ with the boundary conditions $\theta_2 = 0$ and $\overrightarrow{\text{grad}}\ \theta_2 = 0$, at the surface. If θ_2 is only a function of z and r is constant, then we have $0 = K(\partial^2\theta_2/\partial z^2) + r$, which leads to $\theta_2 = -rz^2/2K$. If $r = 1.6 \times 10^{-12}$ or 1.6×10^{-13} cal/cm$^3 \cdot$ s and if z is expressed in kilometers, we determine for the coefficient $r/2K$ the values of 1.7 and 0.17°C/km².

It would be easy, if this were the case, to apply this correction to the subterranean temperatures calculated from surface observations and, in particular, from the measured values of the gradient at the surface. It is clear that for sedimentary rocks, as well as for the metamorphic rocks which are derived from them, we can neglect the effect of radioactivity over the several kilometers where we will study the behavior of subterranean waters. The same is true for intermediate or basic eruptive rocks, which have radioactivities of the same order as above or less.

It is only for granites which are particularly radioactive that the correction term becomes more important and cannot be ignored in the temperature study of a deep tunnel. There are reasons, however, to believe that only the most external parts of certain intrusive bodies have significant radioactivity. We have shown that the correction factor is negative when the surface gradient is known, since the surface gradient is no longer simply proportional to the temperature at some given depth underground. But this does not alter the fact that if the average heat flux at various depths is given, a massive, very radioactive granitic rock will be warmer than its surroundings and will produce a larger superficial gradient.

influences of groundwater

As long as the water which soaks all rocks remains immobile, the water's only effect is modifying the values of the heat capacity and thermal conductivity. It can only be hoped that the measurements of these values has been done with samples saturated with water in the same conditions as would occur in nature. For the thermal conductivity, this is not always the case.

If water moves in the ground, we can characterize its displacement by the *unitary-flow* vector. This vector is represented by **u**, the flux of which across any surface element is equal to the mass of water that crosses the surface per unit time. We assume that this water immediately reaches thermal equilibrium with the rock that it penetrates. The rock gives up to the water a quantity of heat which is equal to the product of **u** times the specific heat of water C_w times the component of the thermal gradient in the direction of **u**. In practice, C_w can often be set equal to unity. The heat equation thus becomes

$$C \ \frac{\partial \theta}{\partial t} = K \Delta^2 \theta - C_w(\mathbf{u} \cdot \overrightarrow{\text{grad}} \ \theta) \tag{3-1}$$

It is also known that the unitary-flow vector of the fluid is connected with its pressure by Darcy's equation:

$$\mathbf{u} = \frac{\sigma}{\eta} \ \overrightarrow{\text{grad}} \ (P + \bar{\omega}gz) \tag{3-2}$$

where P is the fluid pressure and $\bar{\omega}g$ its specific gravity at the point under consideration, σ is normally, i.e., in an anisotropic body, a tensor but can degenerate into a scalar coefficient of permeability if the rock is isotropic. η is the coefficient of the viscosity of water; it varies drastically with the temperature.

It is sometimes necessary to take into account the variations of the specific gravity of water with temperature; these variations are essential factors in water displacements by convective currents. We could try to write a system of unitary equations to describe what have been called hydrothermal phenomena. The difficulty stems from the fact that there is no simple law to describe the variations of the viscosity of water with temperature.

The heat equation written in Eq. (3-1) can be applied without any great modifications to the case of water circulating in a permeable layer of negligible width (such as a fissure) or in a linear channel. The circulation manifests itself by the appearance of heat sources, superficial or linear, whose value is given by the scalar product of the thermal gradient times the flow (but this will not be the unitary flow, but rather the flow per unit width in a linear surface circulation).

As a case in point, and to fix the orders of magnitude, we shall examine the influence on the thermal gradient which would be produced by the absorption of rainwater by the ground. We suppose that rainwater filters in freely from the surface to a phreatic level situated at a depth h beneath which the ground is saturated with water. If we allow a rainwater collection of 31 cm/yr, which is not at all excessive, we find that $u = 10^{-6}$ g/s. Supposing a uniform gradient of 3°C/100 m, it is found that $\partial\theta/\partial t = -6 \times 10^{-10}$°C/s, which is about 0.02°C/yr.

Under the same conditions, let us suppose the phreatic surface is at a depth $h = 100$ m, and suppose that at this depth the temperature is constant (this could be because the sheet of water circulates horizontally and the region where rainwater is absorbed is of a limited area). It is easy to find the steady-state profile for this model, which will be of the form

$$\theta = A(e^{qx/\alpha} - 1) \tag{3-3}$$

which in this case gives $\theta = 1.75(e^{10^{-4}x} - 1)$. The change from the initial linear temperature profile is 0.37°C at middepth, and the gradients are 1°C/57 m at the surface and 1°C/21 m at the base of the perturbed layer. Comparison with the preceding result suggests that a steady state would essentially be attained in several tens of years.

The importance of this rainwater perturbation would completely invalidate all flux measurements conducted at a depth susceptible to the water's infiltration. There would also be other reasons for avoiding this particular range of depths, which could be the site of air circulation and of the condensation or evaporation of vapor. Subterranean seepage from lakes or rivers could have an even more pronounced thermal effect. Measurements of thermal gradient could be designed to show these effects.

In areas of karst topography, where underground caves can have substantial heights, it is impossible to measure the thermal gradient. The temperature there is essentially uniform, and this can be interpreted as the effect of weak air currents which guarantee an efficient transport of heat by convection.

A number of heat-flow measurements have been performed in mines. These mines always constitute water drains and strongly perturb subterranean water systems. We can only hope that gradients measured in mines have not been too strongly perturbed by such influences.

3-1 Convective movements

In a purely conductive system, the steady-state temperature distribution is stable. This means that if such a system is locally perturbed, the perturbation will diffuse, spread, attenuate, and finally disappear on a time scale which is simple to calculate. The steady-state distribution satisfies the heat equation, and the same equation must be satisfied by the perturbation. It is easily verified that the expression

$$\theta = (4\pi\alpha t)^{-3/2}\, e^{-(x^2+y^2+z^2)/4\alpha t} \tag{3-4}$$

is a solution to the heat equation and that its integral extended over all space is equal to unity. For $t = 0$, the expression is zero everywhere except at the origin. Thus, this represents the manner in which a heat contribution released instantaneously at the origin diffuses. Equation (3-4) can also be used to estimate the rate at which the maximum temperature of an extensive perturbation decreases. This process can be characterized by the radius at which the maximum temperature is one-half its initial value, and leads us to represent this maximum temperature in Eq. (3-4) by taking t_0 such that $\rho^2/4\alpha t_0 = \ln (2.0) = 0.67$, which gives $t_0 = \rho^2/2.68\alpha$. Thus, the relative rate of decrease of temperature at the center of the system by conduction is

$$\frac{\partial\theta/\partial t}{\theta} = -\frac{3}{2t_0} = \frac{-4\alpha}{\rho^2} \tag{3-5}$$

On the other hand, if we take into account the presence of water in a porous soil, things can evolve in a different way. We shall study the conditions under which convection currents can develop; they can cause irregularities in the temperature distribution to increase.

Let us first consider the effect of groundwater movement on the transport of heat. The permeability of soil can be expressed either in practical units, grams per square centimeters for a gradient of 1 cm of water per cm, or in darcys σ, the practical unit being $\eta/10$ darcys, where η is the viscosity.

A temperature perturbation θ of the system with respect to the unperturbed system having a thermal gradient m produces a variation of the specific gravity of water given by $\lambda\bar{\omega}g\theta$, where λ is the coefficient of expansion and $\bar{\omega}g$ the specific gravity. This results in an equal variation of the pressure gradient with respect to the unperturbed system. As a result, a certain amount of water is raised and lightened because of its heating (or the inverse), with a flow $q = 10\sigma\lambda\bar{\omega}g\theta/\eta$. The upward motion of the water produces a heating of the surrounding rock at a rate given by $C \; \partial\theta/\partial t = mq = 10m\sigma\lambda\bar{\omega}g\,\theta/\eta$, which leads to a relative rate of change of

$$\frac{1}{\theta} \frac{\partial\theta}{\partial t} = 10m\sigma\lambda\bar{\omega}g/C\eta \qquad (3\text{-}6)$$

It suffices to compare this rate of change caused by the displacement of water by convection with the rate of change calculated earlier [Eq. (3-5)] for the attenuation by conductive diffusion and to see if the perturbations attenuate and disappear, indicating stability of the thermal system, or, in contrast, grow (and if they grow at what rate). The latter case would suggest the possibility of establishing convection currents. We will look for an order of magnitude only, and it would be useless to attempt to describe the convective cells completely. We suppose that other opposing perturbations exist next to the phenomena which we are studying so that the circulation can occur in closed loops.

The value which we found for the relative rate of change of the perturbation by convection can be applied to the maximum temperature. To take account of the fact that within the radius ρ the temperature difference can vary between the maximum value and one-half that maximum, we will reduce the rate of change found for Eq. (3-6) by a factor of $\frac{1}{4}$. We can thus estimate the relative rate of change:

$$\frac{1}{\theta} \frac{\partial\theta}{\partial t} = 7.5m\sigma\lambda\bar{\omega}g/C\eta - 4\alpha/\rho^2 > 0 \qquad (3\text{-}7)$$

This expression must be positive if convection is to be possible.

It is immediately seen that convection will occur much more readily as the thermal perturbation is enlarged. It will be found for each dimension that there is a limiting value of the permeability σ. Convection is aided by a high gradient m. Convection depends especially on the coefficient of expansion λ and the kinematic viscosity $\eta/\bar{\omega}$, both of which vary strongly with temperature. λ, which is 0 at 4°C, increases continuously with the temperature until the critical point is reached and has little variation with pressure. The viscosity decreases greatly as the temperature increases, varying in the ratio of 10:1 between 0 and 100°C, and continues to decrease up to the critical point. The conjugate variation of these two factors makes convection much stronger when the temperature increases.

Table 3-1 indicates the limiting values of the permeability σ (for a gradient $m = 1$°C/30 m), in darcys, above which convection is possible according to Eq. (3-7). For a steeper gradient, these values would be reduced in the ratio $\frac{1}{3.000}$ m.

To provide an idea of the sizes of the above permeability values, for sand of uniform grain size of 0.1 mm, which would constitute an exceptionally permeable rock, we have $\sigma = 1.3 \times 10^{-6}$. However, it is rare that such a permeability is found over a substantial thickness, and we are often concerned with much smaller values. On the other hand, the aforementioned criterion corresponds to a limit for which the rate of development of convection cells would be infinitely small.

Even if in Eq. (3-7), which gives the rate of increase of the thermal perturbation, the first term greatly exceeded the second and, for instance, had twice its value, the temperature perturbation will only double in 53 yr for $\rho = 100$ m, 210 yr for $\rho = 200$ m, 1,320 yr for $\rho = 500$ m, and 5,300 yr for $\rho = 1,000$ m. Unless the permeability is much greater than the limits previously indicated, this increase in thermal perturbation is extremely slow.

The conclusion which we can draw seems to be that at the temperatures which occur at moderate depths there is practically no chance for convection to play a role in a

Table 3-1
Limiting values of the permeability in darcys (for $m = 1$°C/30 m)

θ	50°C	100°C	150°C	200°C
$\rho = $ 100 m	1.2×10^{-6}	3.75×10^{-7}	1.66×10^{-7}	9.3×10^{-8}
$\rho = $ 200 m	3×10^{-7}	9.4×10^{-8}	4.15×10^{-8}	2.42×10^{-8}
$\rho = $ 500 m	4.8×10^{-8}	1.5×10^{-8}	6.65×10^{-9}	3.7×10^{-9}
$\rho = 1,000$ m	1.2×10^{-8}	3.75×10^{-9}	1.66×10^{-9}	9.3×10^{-10}

homogenous permeable region. On the contrary, at the high temperatures often reached in regions with a high thermal gradient, it becomes much more likely that the permeability may be high enough in a sufficiently extended volume to permit convection to become established. In the immediate vicinity of the critical point, convection becomes much easier, and we can consider that it has a good chance of being established when the surrounding rock is permeable over a certain area.

We will see in the following chapter that as soon as the vapor phase intervenes, convection is practically certain.

3-2 A case of intensive circulation

Numerous thermal sources (i.e., sources at a significantly higher temperature than the local average) are known. They may rise with considerable flows through conduits, and it is only in superficial loose soil that they disperse in multiple water filaments. This is proven when one attempts to collect the water. There are many reasons to believe that these are waters which rose from underground as a concentrated current by a direct route or by means of a fissure, fault, karstic duct, etc. We will show, with some schematic models, that in such cases the internal walls warm up quickly, over a certain distance, and that afterward the heat loss experienced by the water in its upward motion becomes negligible.

Descending circulations of colder waters occur under

250°C	300°C	350°C
4.3×10^{-8}	1.86×10^{-8}	5.6×10^{-9}
1.07×10^{-8}	4.65×10^{-9}	1.4×10^{-9}
1.72×10^{-9}	7.5×10^{-10}	2.24×10^{-10}
4.3×10^{-10}	1.86×10^{-10}	5.6×10^{-11}

exactly the same conditions, but they are far less likely to be observed. Certain observations made in tunnels can be noted, however. For example, at Mont Blanc (Italian side), the rock temperature, initially near 30°C, progressively decreased as the tunnel advanced, for some tens of meters, until a fissure in which cold water was circulating was reached. The water naturally flowed into the gallery, but there is obviously no relationship between the amount of water which emptied into the gallery and the flow which initially circulated in the fissure and whose amount remains unknown.

We can imagine the components of a convective circuit as being an intensive descending circulation with a certain amount of horizontal diffusion (in a reservoir where the water can assume the normal temperature of the surroundings at that depth) and then an intensive upward circulation. The difference in density between the water in the ascending and descending paths provides an adequate thermal engine. We could thus have, for example, on the banks of a lake, a thermal source being fed by the lake's water. We could also imagine a circuit composed of a diffuse divided descending water path, with only the upward circulation being intensive.

In some cases we can consider that the temperature distribution, and as a consequence the differences in pressure which make up the circulation, is practically independent of the flow. The amount of water flow will be determined by the head losses of the water, which depend in turn on the shape of the conduit.

In many cases, unless a pure closed convective circuit

Table 3-2

t	1 s	1 h	1 day	1 yr
$K/\sqrt{\pi \alpha t}$	0.0282	4.7×10^{-4}	0.96×10^{-4}	0.503×10^{-5}
$x, \theta = \theta_1/2$	1.35 mm	8 cm	39.5 cm	7.55 m

exists, the differences in pressure gradient caused by the temperature interfere with a normal hydrogeological cycle (i.e., precipitation in going from upper regions to springs at lower elevations) in such a way as to modify the cycle's circulation, and in particular to intensify the thermal currents in conduits.

To make the above statements more precise, we shall calculate, knowing the thermal properties of the appropriate rocks, the way in which conduit walls warm up when in contact with water and thereby diminish the thermal flux that they will absorb. For purposes of calculation, we will assume that the water circulation is established abruptly at an initial time, the rock being at the normal temperature for its depth. This initial phase is not too realistic, but it will give us an idea of the rate at which the conduit walls heat up.

When a large flow circulates in a compact channel, it can be asked if the walls are effectively at the average temperature of the water. We will see that the temperature difference would not be appreciable except for a very short period of time. Thus, it will not be necessary to take account of this difference between the temperature of the walls and of the water.

We shall take as a model of intensive circulation a plane fissure extending for a considerable vertical distance from the surface at which the heating is produced. We can then consider the temperature as solely a function of the distance x from the wall. If the temperature of the wall, initially equal to the rock temperature and taken to be 0°C, is brought to a value θ_1 at $t = 0$, the temperature in the rock is

10 yr	100 yr	1,000 yr
0.159×10^{-5}	0.503×10^{-6}	0.159×10^{-6}
24 m	75.5 m	240 m

given by

$$\theta = \theta_1\left[1 - \mathrm{erf}\left(\frac{x}{2\sqrt{\alpha t}}\right)\right] \tag{3-8}$$

and the flux at the surface is $K\theta_1/\sqrt{\pi\alpha t}$. The factor $K/\sqrt{\pi\alpha t}$ and the distance x at which $\theta = \theta_1/2$ are given as a function of time (for $\alpha = 0.01$ and $K = 0.005$) in Table 3-2.

Consider a fissure rising 100 m to the surface with a temperature difference of 3°C and a flow of $2q$ g/cm of conduit width. The water will reach the surface after a temperature drop of only 1 percent at the end of $(3.7/q^2)$ days. If $q = 1$, which is a flow of 1 liter/s for 5 m of width, this delay would be 3.7 days. For a seepage of 2 g/s · m, which would be difficult to notice, this delay would be 100 yr, which is still inappreciable on a geologic time scale. A fortiori, for a flow of several liters per meter, the delay would be on the order of 1 hour and the heat loss would continue to decrease as the square root of the time. This means the heat loss would be reduced to 1 part in 10,000 at the end of a year.

We could do the analogous calculations for a cylindrical conduit. The wall heating and the reduction of the heat losses at first would be the same as for a fissure with the same surface area and afterwards a little slower. It is not necessary to make these calculations, which are a bit more difficult than for a plane fissure.

It should be remembered that whatever the form of the conduit in which an intensive circulation takes place, several hours or several days are sufficient for the walls to come into thermal equilibrium with the water so that conduction losses become negligible.

Conversely, if we knew the shape of the fissure and the law with which the rock temperature varied with distance, we could try to calculate the time when the circulation became established (like the case earlier cited for the Mont Blanc tunnel).

When the search for hot water is made by driving adits, as has been done many times in the Pyrenees, the apparent rock temperature is often a guide. But we cannot exclude, a priori, the possibility that the temperature distribution may correspond to a steady-state system between the thermal water filaments and the surface (or possibly cold-water circulations).

3-3   Illustration of Hammam Meskoutine

This Algerian locality, well known for its thermal sources and the concretions which these sources have produced, will furnish us with a remarkable illustration of the stabilization of a conduit by convection.

A small plain is found there (see Fig. 3-1) having hundreds of limestone formations of about the same height (around 3 m). Each of these formations corresponds to a thermal conduit from which water once emerged at a temperature near 100°C and has produced abundant concretions, raising the conduit up to a height of 3 m above ground level. When the conduit reaches this height, it ceases to function and another conduit develops next to it. When the conduit is functioning, its walls are heated and the entire water column is at 100°C for which the specific volume of water is 1.0434.

On the other hand, in a fissure without water circulation the temperature should vary progressively between 100°C at the level of the subterranean aquifer and an average 20°C found at the surface. The average specific volume in this temperature range is 1.0193, and thus the pressure difference at the base of the source is $0.024h$ (in meters of water). If this difference is capable of overcoming the additional height of 3 m due to the concretions, then $0.024h \geq 3$ which means $h \geq 125$ m.

We can thus estimate the depth of the aquifer where water at 100°C is found and estimate the local gradient at 1°C/1.6 m. It would take us too far afield to discuss the underground thermal system at length, and besides it is still poorly understood.

For a number of hot springs, there is the same effect of stabilization of the acting spring, as compared with incipient ones, with no discharge, which remain cold. This pre-

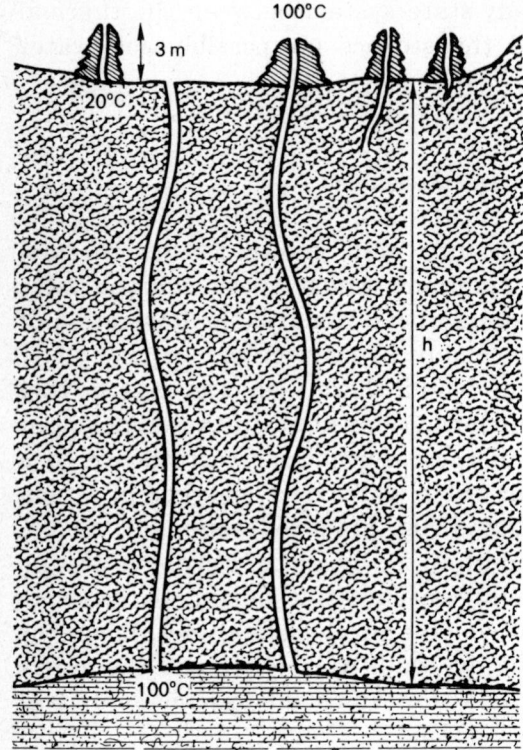

FIGURE 3-1 Diagram of the convection at Hammam Meskoutine (Algeria). The height of the concretions on successive conduits reaches 3 m and the height *h* for which there is an equilibrium between the conduit entirely at a temperature of 100°C and a potential conduit where the temperature of the immobile water varies progressively from 100°C to 20°C can be calculated.

vents loss of thermal water toward other springs and facilitates their exploitation.

3-4 Thermal effects in a drill hole

A drill hole consists of a cylindrical space filled with water (or mud) whose boring has given rise to strong thermal perturbations. It is necessary to understand these effects if we wish to use drill holes for underground temperature measurements.

In this section, we can distinguish between three phenomena: the perturbations resulting from drilling operations, the manipulations which can be performed to modify the drill hole's thermal regime, and the regime attained at the end of a waiting period.

In the course of drilling, the drill bit absorbs a significant amount of mechanical energy which is mainly transformed into heat. Water (or mud) is pumped into the drill stem which cools the bit and then rises between the stem and the surrounding rock. It is stored in mud tanks, where it generally does not have time to cool completely before being reused. It is normally cooler than the rock in the deeper parts of the drill hole, which it cools. One could propose to measure the temperature at the farthest point reached by the drill after letting a waiting period expire, such as rest on Sunday. This temperature would be representative of the temperature of the surrounding rock if the thermal perturbation at the bottom of the hole, during the brief and finite drilling time, had time to dissipate during the waiting period. Unfortunately, it is not generally possible to lower a thermometer suspended on a cable to the bottom of the hole, which is blocked by caved-in debris or sedimented mud. To reach the bottom of the hole, it would be necessary to reestablish circulation in the stem, and that would perturb the temperature distribution.

In the course of the boring, the surrounding rock will be cooled in the lower regions and heated in the upper regions. A given part of the drill hole will thus be successively cooled and then heated while the drilling progresses.

When a geothermal bed is drilled, it is customary practice to cool the mud at the surface by means of an aerated tower. Even if this cooling is small, 10°C for example, it prevents a progressive heating of the mud which would be bothersome if its temperature approached 100°C. In such a case, the walls are only heated by the drilling operation very near the surface, and over most of the height the walls are cooled, sometimes very strongly.

The rhythm of work in a drilling operation is too irregular to practically permit the calculation of the penetration of the thermal perturbation into the walls and the evolution of the thermal flux at a particular depth, as would be theoretically possible. We can only make a rough estimate of the order of magnitude of this penetration as a function of the drilling period.

Once the drilling is finished, it is possible to obtain certain hydrogeological data from the thermal profile measurement. For example, if an aquifer is cut by a drill hole, we can produce a uniform temperature along the hole by circulating the mud through pipes lowered to the bottom of the hole. After a waiting period, a thermal profile will indicate the entry level of the water at the temperature of the surrounding rock. It will appear warmer in the lower part of the hole and cooler in the upper part.

If there are indications of gas, they can be located (they give rise to endothermic expansions) by lowering the water level by pumping and determining the depth of the characteristic temperature drop on a thermal profile due to gas.

Finally, if we are trying to cement a collar in a drill hole, the best way to determine the height reached by the cement is to observe on a thermal profile the heat produced by the setting of the cement.

In these operations, the thermal measurements have well-determined objectives and give us poor information at best about the temperature of the surrounding rock. Given the above objectives, thermometers are sometimes not standardized precisely. In general, these measurements are of very short duration, and the supplemental perturbation which they bring to the wall temperature is small compared to the perturbation caused by the drilling.

Once the drill hole is left undisturbed, it frequently happens, especially when sedimentary formations are traversed by the drill, that permeable layers are encountered which are not in pressure equilibrium with one another. In these circumstances a mixing circulation can be established in the drill hole and eventually outside the tubing, if it is not cemented, running, for example, from the deepest permeable layer toward the shallower permeable levels. It was found in the previous case considered that, in the section of the drill hole where the mixing circulation occurs, the walls of the hole are reheated after a sufficient time to the temperature of the deepest aquifer. The opposite process can also occur: in the southern part of France a drill hole registered a temperature essentially the same as at the surface in penetrating a sedimentary layer. It then encountered a discontinuity and a normal thermal profile in the crystalline basement. From such a thermal profile it is quite easy to recognize the sections of the drill hole in which the water circulates. It can also happen that the undisturbed portions of the thermal profile are aligned and give the impression of furnishing a true thermal profile of the ground; however, it must always be considered that smaller circulations of water could still exist.

Implicitly, it has been assumed that thermal profiles are measured long enough after the end of drilling operations so that spontaneous circulation of water has been fully developed. At the same time, the thermal perturbations arising from the drilling operations must have had time to dissipate from the walls of the hole. When the purpose of the drill hole is to measure a completely representative thermal profile, it is often necessary to make at least an approximate calculation of the time necessary for such perturbations to have disappeared.

In Chap. 1 it was seen that in cylindrical coordinates the solution $\theta = (1/4\alpha t)e^{-x^2/4\alpha t}$ represents the diffusion of a thermal perturbation produced along the z axis at the time $t = 0$. Such is not the case for a drill hole when it is left at rest. The thermal perturbation, whatever its sign, has penetrated into the walls of the hole to a depth which ought to increase as the square or cube root of the duration of the drilling operation. For example, we might take the values indicated earlier for the depth of penetration of half the maximum temperature drop as a function of time. Taking the radius of the hole into account as well, we can estimate the radius at which the thermal perturbation coming from an axial source would reach the same distance. For example, at the end of a drilling period of several weeks, say, we estimate that the halfway point of the perturbation should be found at a radius of 2 m, from which we get $\ln 2 = x^2/4\alpha t = 200^2/0.04t$, and thus $t = 1.4 \times 10^6$ s = 16 days. We deduce that 16 days later the perturbation will be reduced by half, 32 days later it will be a fourth and that it would be necessary to wait 144 days in order that the perturbation be reduced to less than 10 percent of its initial value.

Of course, it is difficult to estimate the thermal perturbation distance in the walls of a drill hole. We can hope that different perturbations will partly compensate for one another, but that is not a reason to underestimate the waiting time necessary so that the measured thermal profile will be perfectly representative.

For the geothermal drill holes in New Zealand, where the shaft walls are strongly refrigerated during the drilling, it is current practice to wait 3 months before measuring the thermal profile. What has just been described shows that this delay is not at all excessive. Unfortunately, it is to be feared that a number of the published values of heat flow are based on thermal profiles measured without a sufficient waiting period.

There is one last consideration that we should mention, although it is rarely taken into account. In the water column inside the drill hole there exists a certain thermal gradient. Does this produce convection currents which would agitate the water and tend to regularize its temperature? This is a problem in hydrodynamics which is outside the scope of this book, but it has been indicated that for the thermal gradients normally encountered, convection will only occur for large diameters (50 cm and more). This is one reason (others are chiefly economic) to make exploratory geothermal drill holes of small diameters, even if commercial holes must be of a larger diameter.

3-5 Example of a sheet circulation

We have successively seen the convection which can result from a movement of water in a permeable volume and the effect of an intensive circulation in a limited channel. The following problem, which is encountered in the study of certain geothermal regions, will furnish an example of the effect of water circulation along a given layer. We suppose that at a certain depth there is a sheet circulation of water in a layer whose thickness we will neglect. The given flow is q per unit width. Now, suppose that there is a localized underground thermal anomaly such as would be produced by a heat source. The water flows will cool the ground and heat it downstream. Thus, the thermal anomaly will appear displaced with respect to where it would be in the absence of circulation. We wish to calculate the distribution of the surface gradient or any other aspect of the temperature distribution once the steady state is reached.

If the surface of the ground is horizontal, we can satisfy the condition $\theta = 0$ (taken for a constant value) by adding fictitious heat sources, symmetrical and with opposite sign, to those sources which exist underground. In the absence of a circulation of water, the combination of these heat sources (including a source at infinity to take the gradient into account) determines a temperature distribution θ'.

In the presence of the water circulation, the temperature distribution becomes θ. Along the water's path, there is a new source of heat $q\, d\theta/ds$ at each point which determines a temperature distribution

$$\Delta\theta = \frac{q}{2\pi K} \frac{d\theta}{ds} \ln\,(r)\, ds$$

and the temperature change is due to the sum of these perturbations,

$$\theta - \theta' = \frac{q}{2\pi K} \int \frac{d\theta}{ds} \ln\,(r)\, ds$$

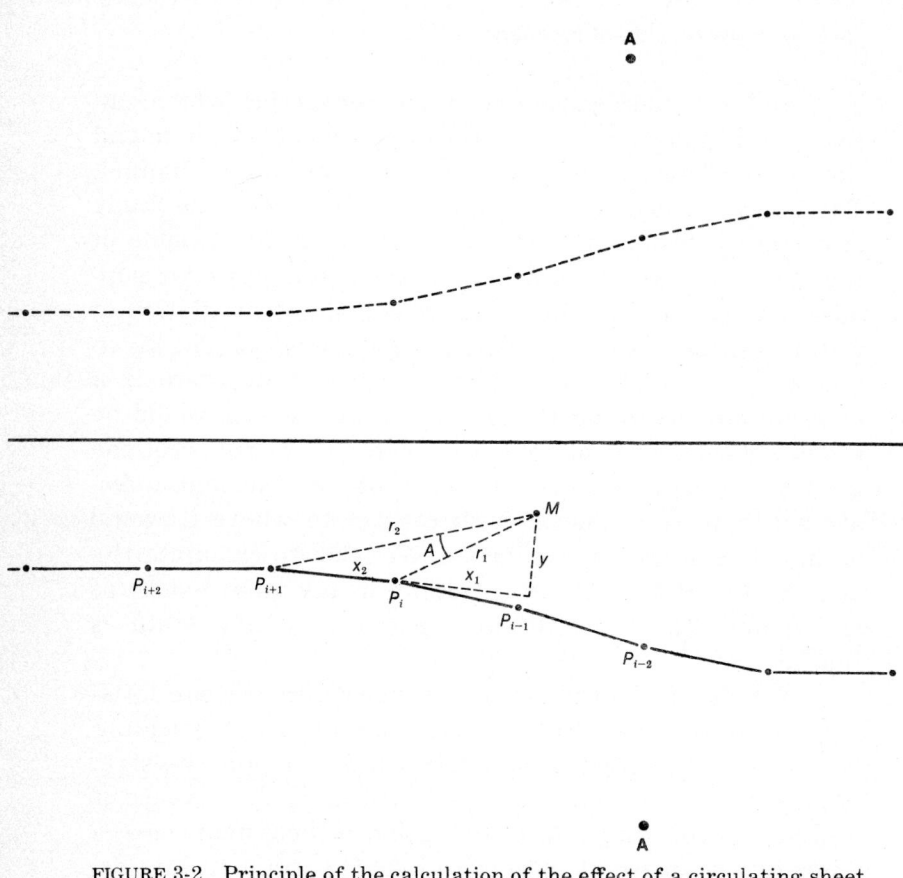

FIGURE 3-2 Principle of the calculation of the effect of a circulating sheet
of water following the path P_i, $P_i + _1$, ... on the temperature
distribution defined by a gradient, localized heat sources A,
and the condition of a constant temperature on a horizontal
surface.

where r is the distance from the point where we are
calculating the temperature to the point along the water's
trajectory where we perform the integration (do not forget
to also carry the integration over the fictitious symmetrical
path). To lead to a solution by a numerical method of approximation, we will take the final temperature at a limited
number of points P_i along the water's path as unknowns.
In addition, we will assume that from one point to another
the temperature variation is linear (see Fig. 3-2). This

allows us to write, between two of these points, $d\theta/ds = (\theta_{i+1} - \theta_i)/s$ and to write this constant outside of the integral sign. As for $\int_{P_i}^{P_{i+1}} \ln(r) \, ds$, this is a simple calculation. If x_1 and x_2 are the projections of $P_i M$ and $P_{i+1} M$ on the line $P_i P_{i+1}$, and y is the distance from M to this line, A the angle $P_i M P_{i+1}$, and r_1 and r_2 the distances $P_i M$ and $P_{i+1} M$ respectively, then

$$\int \ln(r) \, ds = x_2 \ln(r_2) - x_1 \ln(r_1) - (x_2 - x_1) + yA$$

It can be verified that this expression remains finite at P_i and P_{i+1} and that, if $y = 0$, this reduces to $x \ln(x) - x$. Of course only a computer allows us to solve this system of linear equations by equating each of the unknown temperatures to the sum of the initial temperature values at the considered point and the perturbations brought in at that point by the successive segments along the water's path.

The calculation is possible for any form of the water path resembling a polygon. The formulas are simplified if the water path is horizontal.

changes of state
of groundwater

4-1 Freezing

Water is transformed into ice at a temperature of 0°C; this transformation temperature changes very little with pressure. The freezing of 1 g of water liberates a latent heat of about 80 cal. The temperature at which a saline solution deposits crystals—in general, pure ice—can be appreciably lowered depending on the solution's concentration. The specific heat of ice is about half that of water, and the thermal conductivity of a loose soil saturated with water that has frozen is multiplied by a factor of 2 or 3. In addition remember that the density of ice is 0.9, and hence freezing is accompanied by an increase in volume. Actually, we will see that this phenomenon plays only a minor role.

To a first approximation, which would be valid if the groundwater content was infinitely small, what we have seen in Chaps. 1 and 2 shows how and where a temperature of 0°C can be reached: on the one hand, in the winter in a surface zone where the annual temperature variations penetrate, and on the other hand, if the mean annual temperature is less than 0°C. In the latter case there may be an underground frozen region, with an overlying layer that may thaw during the summer. This is what is called *permafrost*, and it covers vast areas in the north of Siberia and Canada. Permafrost can be found to depths of several hundred meters, which implies that its origins are very old (beginning of the Quaternary).

Observation confirms these general indications, which can be completed by saying that in consequence of climatic fluctuations from one year to the next, it can happen that the surface which thaws above the permafrost during the summer may not completely refreeze the following winter, or that the frozen winter surface does not completely thaw the following summer. Thus, alternate frozen and thawed layers can exist above the permafrost. Naturally, these layers are not continuous.

To go a bit farther, we should remember that the water content of the surface areas affected by these phenomena is in general considerable. Their freezing therefore absorbs appreciable heat. From this, what has been called a *conduction barrier* results. All incident heat flux, whatever its direction, is absorbed by fusion or freezing across the zero isotherm and is not propagated further down. Its only effect is to displace the freezing boundary, and while this surface is being shifted, the flux and gradient values on either side of the barrier are very different. If $\bar{\omega}$ is the porosity, u the displacement velocity of the freezing boundary, L the latent heat (80 cal), and K_W and K_I the conduction coefficients for thawed and frozen ground, respectively, the difference between the thermal gradients is given by

$$K_W \left(\frac{\partial \theta}{\partial x}\right)_W - K_I \left(\frac{\partial \theta}{\partial x}\right)_I = L\bar{\omega}u \qquad (4\text{-}1)$$

It immediately becomes apparent that the calculations of Chap. 1 concerning the propagation of periodic temperature variations into the earth's surface are no longer valid. Nonetheless, the general behavior of the phenomenon stays the same with at most a decreased depth to which the temperature variations propagate.

A more important perturbation results from the fact that water inside fine capillaries does not freeze until temperatures are well below 0°C. This fact is easily observable in the freezing of a surface layer of mud. The mud does not freeze altogether in a body, but thin sheets of pure ice appear between which the mud desiccates and becomes powdery. At a few degrees below 0°C, there is a thermodynamic disequilibrium between the water which is still in liquid form in the capillaries and the thin lenses of ice which have formed near by. The water tends to move from the pores toward the ice crystals, which enlarge and are capable of compressing the surrounding ground.

This phenomenon has been especially studied for roads which can swell if ice crystals forming in inevitable cracks are fed by capillarity from below; they may grow to large sizes, resulting in a swelling which can become quite large (several centimeters or tens of centimeters). At the beginning of the thaw, ice melts on the spot, and the pavement becomes flooded with an excess of water. A small load can severely degrade roads and from this comes the institution of the "thaw barrier"—temporarily forbidding certain roads to heavy trucks. However, the real solution lies in preventing the ascension of water by capillarity to roadways which are likely to freeze. This can be achieved by interposing a barrier layer under the road consisting of cobbles without any fine filler.

Another consequence of the way in which ice can form in the ground is "cryoturbation." This involves special types of ground deformations caused by repetitive alternations of freezing and thawing. Ice lenses can form under stones (which conduct heat better than the surrounding soil), and this can lift the stones up. Doubtless this is the way that the *soil polygons* in arctic zones are produced. These polygons are also sometimes observed in mountains such as the Alps, whenever loose soil forms a horizontal surface above 2,700 m.

For permafrost, this unusual type of ice formation can also play an important role. It explains the underground formation of lenses of pure ice either horizontally, forming hydrolaccoliths, pingos, etc., or vertically, forming wedges in cracks and crevices whose traces (after thawing, they are filled by the fall of surface gravel) are frequently visible in ancient alluvial deposits. These cracks appear to form large polygons with dimensions of 10 to 20 m.

Let us add that the circulation of subterranean waters, either in the unfrozen surface layer (permafrost is totally impermeable) or through gaps in the permafrost, plays an essential role. Lakes and watercourses also introduce per-

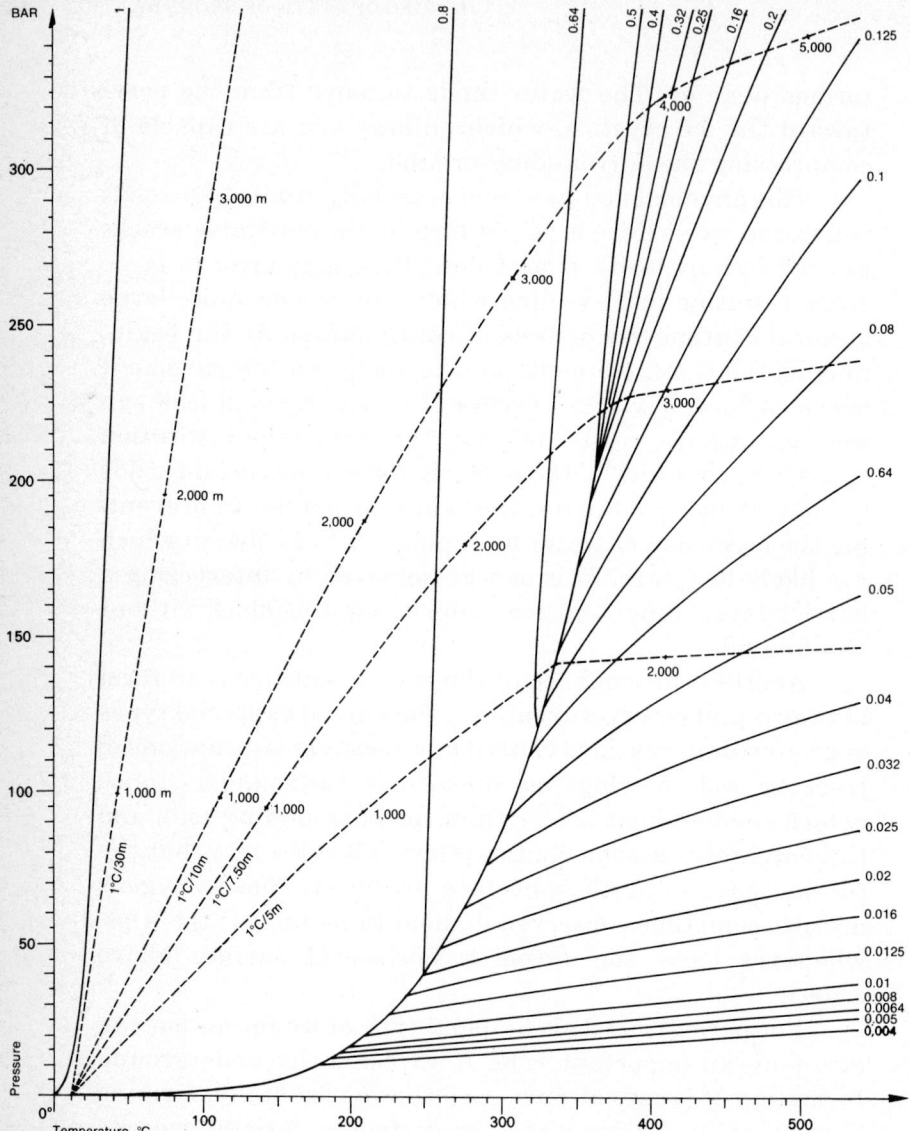

FIGURE 4-1 Graph of water density as a function of temperature and pressure. The dashed lines indicate the states as a function of the depth for a fluid phase in static equilibrium with a temperature of 11°C at the phreatic level and the different values of the thermal gradient, supposed uniform, taking account of the water density. Note that the critical state of water is reached for a gradient of about 1°C/7.5 m at a depth of 2,700 m. For a higher gradient, the hypothesis of a uniform thermal gradient becomes incompatible with the hypothesis of a static fluid equilibrium, this equilibrium being unstable.

turbations which can break up the continuity of the permafrost.

The existence of permafrost presents particular engineering problems for highways, homes, and pipelines which will not be discussed here.[1]

4-2 Vaporization

The transition of water to a gaseous form (steam) occurs at a temperature which is a function of pressure (Fig. 4-1). This transition absorbs a considerable latent heat, which appreciably diminishes when the temperature increases. The density of steam, like all gases, diminishes slightly when the temperature increases and increases greatly with pressure. At the critical point ($T = 375°C$, pressure = 221.06 bars), the specific volume of steam becomes equal to that of liquid water (3.1975), the latent heat falls to zero, and there are no longer any differences between the two phases. For temperatures and pressures with higher values than the critical point (supracritical domain), there exists only a single fluid phase whose density varies in a continuous manner. There are tables which furnish for water as well as for steam all the pertinent characteristics as functions of temperature and pressure, in particular the specific heat. The viscosity values appear to be the least well known in the supracritical domain (see Fig. 4-1)

If the water contains dissolved gas, it will accumulate in the vapor phase, where the total pressure will be the sum of the partial pressures of the gases and the water vapor. The vapor will exchange matter with the liquid until the pressure of the latter is in equilibrium with the total pressure of the gaseous phase.

On the other hand, dissolved salts are distributed very unequally between the two phases, nearly all being found in

---

[1]A. Cailleux and G. Taylor, Cryopédologie, étude des sols gelés, no. 1203, Act. Sc. and Industrielle, Hermann, Paris, 1954; S. W. Muller, "Permafrost, or Permanently Frozen Ground and Related Engineering Problems," 231 pages, Edwards, Ann Arbor, Mich., 1947.

the liquid. The presence of salt can have the effect of increasing the critical point. This means that above the critical temperature of pure water there can be two phases, one with a high salt content and thus more dense, and the other phase low in salt content (the water content in relation to the volume being about the same). Depending on the salts and provided that both solutions are saturated (presence of an excess of solid salt), the distinction between the two phases can continue whatever the temperature and pressure (such as with NaCl), or it can end at a critical point beyond which there is only one unique phase (such as the case of SiO_2). For less concentrated solutions, there is only a slight rise in the critical point.

Having recalled these properties, under what conditions can we find water in the ground in the form of steam?

First, of course, it is necessary to consider the neighborhood of the earth's surface. In the atmosphere, the partial pressure of water varies from place to place. It particularly varies with the temperature, the atmospheric pressure changing very little. In general, there exists, above the phreatic level, a certain fringe region in whose pores air can circulate. Depending on the temperature changes, the water vapor can condense or, on the contrary, the region can desiccate. These transfers, and the heat exchanges which they imply, must be considered as part of the complex surface phenomena which contribute to the determination of the average ground temperature. We shall not study these surface phenomena further.

4-3 Geysers

A given hydrogeologic situation can involve the ascending of waters from the depths through faults or karstic conduits. We have already seen that such springs can reach the surface with hardly any heat loss if their flows are large enough.

A priori, nothing prevents subterranean waters having temperatures exceeding 100°C from reaching the surface.

The adiabatic transport (i.e., a transport without heat exchanges, which will be the case once a steady-state system is achieved) of such waters up to atmospheric pressure necessarily entails a partial vaporization in a proportion x, which is easy to calculate. If $\theta > 100°C$ was the water source's initial temperature, then $x = (\theta - 100)/540$, 540 cal being the heat of vaporization at 100°C. But, although this can happen, we will not necessarily have two regular flows, one of water at 100°C and the other of steam at the same temperature. Especially if the flow is large and the conduit by which the water reaches the surface has a suitable form, boiling can have a discontinuous nature and produce periodic expulsions of water and steam, which are characteristics of geysers.

When a geyser is inactive, the conduit is filled with water which, at each depth, is at less than the boiling point for the local pressure (100°C at the surface, 120°C at 10 m, 135°C at 20 m, etc.). But with the arrival of warmer underground waters, this water column will progressively heat up. Finally, a moment will arrive when boiling will start somewhere and that will lighten the column thus diminishing the pressure sufficiently so that boiling will become generalized and lead to an expulsion of some water to the outside. The steam will escape, and, to finish the account, the water which has not been expelled and which has been cooled by a partial vaporization will constitute a new water column in thermal equilibrium in the conduit. Numerous complications can be imagined, for example, lateral pockets in which either hot water or pressurized steam can accumulate. The heat capacity of the walls can play a role in the regularization. However, we are still a long way from explaining all the singularities of geysers. For example, why does a geyser's eruption come a few minutes after soap is thrown in the conduit? Nevertheless, it can be considered as an established fact that a geyser is a manifestation of the emergence of subterranean waters which are in a liquid state above 100°C underground and which circulate in concentrated enough conduits to have heated the walls in such a way that the water's heat losses are reduced.

4-4 Vaporization at depth

Leaving aside the case of intensive circulation, we now imagine the situation resulting from a uniform geothermal gradient in a somewhat permeable region. Below the phreatic level, the pressure due to the hydrostatic head of water rapidly increases. There can exist complex hydrogeological situations in which the head, relative to the different water layers separated by impermeable layers of rock, increases as we go to deeper and deeper pools of water. Sometimes the subterranean fluid pressure tends to approach what is called the *lithostatic pressure*. This is the pressure caused by the weight of overburden, and this limit cannot be exceeded by the fluid pressure, since the ground would then tend to rise and produce fissures, which would allow the water to escape. The critical pressure of water is reached, that is, is equal to the lithostatic pressure, at a depth of 800 or 900 m. However, at these depths, the temperature will almost always be less than the critical temperature (375°C, that is, a gradient less than 1°C/2.5 m), and so the water remains liquid. In a region where the fluid pressure tends to equal the lithostatic pressure, however, unusual local fissuring can allow a transport of water to the surface and cause a drop in fluid pressure which, if the temperature is sufficient, can lead to vaporization. Such is the situation at the Geysers (California), which is a geothermal field, situated in the middle of the Franciscan formation in which the fluid pressure nearly equals the lithostatic pressure.[1] On the other hand, the Geysers correspond to a zone with a high thermal gradient which could be in contact with a hypothetical underground batholith; it is also a zone with a low fluid pressure. We cannot tell if this vapor occurrence is a consequence of local heating or of fissuring which would have reduced the pressure, or both.

Leaving these complications aside, let us concentrate for the moment on the simple case where, the ground being somewhat permeable, the impregnating water is at a hydro-

[1]K. A. Frederix and F. Berry, High Fluid Potentials in California Coast Ranges and their Tectonic Significance, *AAPG Bull.*, vol. 57, no. 7, pp. 1219–1249, July 1973.

static pressure in equilibrium with the water at the surface or at the phreatic level, which is equal to the hydraulic pressure produced by a continuous column of water, the summit of which coincides with the phreatic level.

Assuming that the thermal gradient, which we will take as a constant to simplify matters, is known, it is easy to calculate, as a function of depth, the hydraulic pressure and temperature at the same time. We can calculate the hydraulic pressure by taking for each depth the actual density as a function of the temperature and the pressure (see Fig. 4-1). For usual values of the gradient, the values obtained exceed the critical pressure for water at a much lower temperature than the critical temperature (i.e., there cannot be any vaporization).

In order for this not to be the case, the gradient, supposed uniform, must reach 1°C/7.5 m. The critical point is then reached at a depth of 2,700 m (supposing the phreatic level has a temperature of 11°C). For a higher gradient, we would expect to find a layer of liquid water on top of vapor. It is clear that this is an unstable situation. Even if the rock is only slightly permeable, it is inevitable that water will descend in some places and steam will rise in others. Where the water descends, it encounters hotter rocks and cools them while being partially vaporized. Where the steam ascends, it meets cooler rocks and partially condenses while heating the rocks, and the pressure drop brings forth more steam from underground. Our implicit assumption of a regular thermal stratification fails. In fact, there will be warmer zones in which steam rises, and cooler zones where water descends.

Such a convective system can be established for a mean gradient such that the critical point and the vaporization level are not reached. In the supracritical domain, but near the critical point, the density varies greatly with the temperature and the onset of convection, even if the rock has low permeability, is very likely. Such a convective action must progressively increase. The cooler descending branch will remain in the supracritical domain, but the ascending branch can reach and exceed the critical point. Convection will then be composed of a vapor phase, with a finite density difference with respect to water and a certain latent heat. This vapor phase rises easily, and on condensing heats the rock which permits the continuation of the cycle.

We can thus expect that the convection currents, the onset of which are inevitable if the gradient is such that the critical conditions are fulfilled, will take two forms, one cooler descending water currents which remain supracritical (i.e., in a dense form), and the other of ascending steam currents which progressively attain a higher and higher elevation. The low density of the vapor with respect to water determines the pressure differences which constitute the driving force for such a circulation.

It is hardly possible to tell what the transverse dimensions of the convection cells were initially, but, if they were small and numerous, it is to be expected that certain ones among them would develop more rapidly than others, and for these the convective action risks being interrupted by a pressure drop underground caused by larger cells. We can predict an evolution toward a small number of important ascending cells, each of which must produce a substantial flow of vapor.

The possibility that a supracritical domain still reigns far underground is not excluded. This means that the relatively cold descending water, which comes into contact with rocks progressively heated by the high local geothermal flux, passes in a continuous manner, without change of state, into ascending currents which are brought about by a pressure drop, and which are individualized as vapor currents making their way above the level where the critical pressure is found.

In the region where the water descends, there can easily be temperature fluctuations, depending on whether the circulation is more or less rapid, but the density variations resulting from such fluctuations are small. Thus it is unlikely that these fluctuations in the rate of circulation tend to be amplified. Rather, we must expect an almost uniform descent over a certain area.

These considerations lead us to predict that if locally

the geothermal flux exceeds 4 or 5 $\mu cal/cm^2 \cdot s$ (we consider only the surfaces of the continents, because in the ocean depths the pressure already exceeds the critical pressure and the conductive flux can reach greater values) and if the rocks have a certain permeability, even though it may be small and irregular, heat transfer by conduction, which implies a regularization of the thermal gradient, will not act alone. Eventually a relatively restricted number of zones of rising steam must appear in which the pressure gradients corresponding to the density are low enough so that the flows reach high values, possibly with considerable head losses. As the rising steam reaches cooler layers, it condenses while heating these layers, and the heated zone thus tends to rise higher and higher. The condensed water and the cooler groundwater tend to descend along the margins of the steam currents because it is there that the pressure gradient is smallest. These cooler waters tend to give the sides of the steam plume an abrupt nature with strong horizontal temperature gradients. The entire water balance can be put in equilibrium by a steady descent of groundwater outside the ascending steam plumes, but direct evidence for such a descent is lacking.

In the process which has just been described, an essential part—the steady passage of descending supracritical water to ascending currents which individualize themselves higher in the steam plume—has escaped observation till now because boreholes have never been drilled to such a great depth (it would be necessary to go to 3,000 m) in abnormally warm regions. Hence, the origin of this abnormal heating remains largely hypothetical. We can imagine the injection of eruptive rocks, for example, a granitic batholith, the cooling of which could supply a large heat flux for a considerable time.

But under these circumstances, petrology leads us to suppose that crystallization of the granite would liberate a certain quantity of volatile substances which could have been dissolved in the molten magma. Water is the most abundant of these substances, but there can be many others (CO_2, Cl, sulfur in different forms, B, etc.). A portion of these volatile substances could be found in the ascending vapor phase. But, on the other hand, the vapor, especially in the supracritical domain, can, while circulating in contact with rocks, produce diverse alterations in them and leach certain elements (in particular, steam will dissolve silica, which is always present).

The "juvenile" or *phreatic* origins of the steam found in geothermal pools has been the subject of much discussion. There is nothing to exclude the possibility that part of the steam, and especially the volatile impurities, are of a juvenile origin (i.e., they originated directly from a magma in the course of its consolidation and cooling). But that can only represent a small amount of the steam. The cooling of each gram of granite will supply 25 to 40 cal from the latent heat of crystallization and, for a cooling of 200°C, about another 40 cal; but the magma could scarcely contain as much as 10 percent in volatile substances even if the magma had been entirely molten; if the fusion was only partial, which would be very likely, the total content of volatile substances would be much less. From these numbers, which are only orders of magnitude, the fact emerges that the volatile products given off by a granite would not be able by themselves to transport the amount of heat liberated by the granite's cooling. This makes it very probable that the granite will produce a high thermal gradient leading to, besides an enhanced heat transport by conduction, the establishment of a substantial convective movement of phreatic waters.

4-5 Evolution of a convective system near the earth's surface

The behavior of the ascending vapor plumes which transport heat in excess of that carried by conduction (about four times the normal flux) can take quite different forms depending on the structures encountered.

Even if the paths along which the steam reaches the surface are convoluted, it is to be expected that its ascent is very rapid, since its low density entails a nearly negligible pressure gradient near the surface. Thus the removal of the excess subterranean heat will be rapid. On the other hand, neighboring waters tend to invade the geothermal site, especially at its base, and steam condenses in rocks which are cooled in this way. We can imagine that a final situation is thus established in which the terrain is entirely invaded by water, but which at each level is at the boiling temperature under the local pressures. When that happens, pressure equilibrium is substantially realized and the waters can remain practically immobile. However, as long as a vapor "bubble" exists, it must tend to rise because of its low density, even through regions of low permeability.

Research done in New Zealand has shown that in geothermal sites, this temperature distribution is essentially the one that is found with the rocks containing water in the liquid state. Many measured temperature profiles are typical in this respect. The geothermal zones where these conditions take place are of a limited area and are separated by very sharp boundaries from surrounding regions in which the thermal gradients are high but regular.

In geothermal sites below an underground depth of about 500 m, the temperature ceases to increase, due to the invasion of neighboring cooler waters. Such a temperature distribution is established even when the terrain (as is the case in New Zealand) consists of an alternation of layers with very different porosities and permeabilities. The permeable strata (it may be chiefly fracture permeability) constitute reservoirs from which hot water can be extracted.

4-6 Underground steam accumulation

A very different case exists when there is an impermeable cover, in any form which will assure closure (i.e., a reservoir with its higher rim at a lower level than the summit). The reservoir can be of any geometry, and, for example, at Larderello (Italy), it consists of an assemblage of karstic voids in the Triassic and Liassic limestones overlying a series of horsts in a crystalline basement, covered by an impermeable layer of exotic Oligocene "argile scagliose."

The result of such a covering is that steam accumulates in the reservoir. When there is a local cooling and condensation, water flows to the bottom of the system and is replaced by steam of subterranean origins. This steam remains in a saturated condition (at Larderello, 240°C at 35 bars), and its density is much lower than that of water. The pressure and temperature vary only slowly with the depth. The walls of fissures or steam-filled spaces are at the same temperature as the steam, but this temperature is quite different from the temperature of the rocks included between the fissures. A sounding at Larderello has shown an approximately regular thermal profile with a gradient of the order of 1°C/8 m, disturbed only by a few temperature anomalies in the Trias, probably caused by neighboring steam-filled fissures with which communication has not been established. We should keep this evidence in mind when we examine the age of a geothermal site.

4-7 The age and the origin of geothermal beds

The interpretation which we have given to the genesis of thermal layers implies that these generating processes may no longer be active today. This is evident for those processes where a rock saturated with water is found at the boiling point as determined by the local pressure. The accumulation of steam under an impermeable cover, which provides a proper closure, should persist for a certain time after the end of the arrival of the subterranean steam which gave rise to the accumulation.

Certain prospectors, in New Zealand for example, hoped to sink holes that would reach the subterranean source that was supposed to supply heat to the geothermal sites. Unfortunately, this hope was not realized. How should we react to this?

What is the cause of the subterranean heat which produces exploitable geothermal sites as a superficial manifestation? We could immediately think of volcanic phenomena, but how would they act? We could invoke the formation of a laccolith or a batholith which could have gradually cooled over a very long time. But we should also consider the possibility of a network of dikes which are individually very thin and thus susceptible to rapid cooling, or even to the release of steam. In any case, we know by numerous observations of volcanic eruptions that eruptions are essentially irregular in time. Active periods can last for several days or weeks and be separated by years of apparent calm. It is for this reason, rather than because of technological difficulties, that the direct utilization of volcanic energy has not been envisioned. Volcanic activity could perfectly well be the primal cause of exploitable geothermal sites. The heat flow is regularized by the enormous thermal capacity of the underground rock mass. However, we cannot determine whether this regularization occurred as deep as the high-temperature rocks. Such would be the case if the deep source were a batholith or a large laccolith. If the heat flux at high temperatures, which is expressed by the formation of subterranean vapor plumes, participated in the irregularity of volcanic phenomena, the regularization resulting from accumulation of heat by the rock mass may occur only at the level of the geothermal sites themselves.

In any case, we can easily determine the age of a geothermal site. In effect, if a geothermal site is not supplied with heat, simple conduction must lead to a progressive temperature decline and thus to the site's degradation. This is true even without taking account of possible convection or losses through surface manifestations such as geysers, steam discharges, mud pots, etc. It is easy to calculate the conductive cooling for a realistic model, and the study of this cooling will allow us to estimate an order of magnitude for the maximum age (i.e., since the period in the course of which the geothermal site must, however formed, at least have experienced a rejuvenation by a new supply of heat).

To start the calculation, we will consider a geothermal site indefinite in the horizontal plane and characterized by a thermal profile as a function of depth. In this profile, we must distinguish between the normal permanent geother-

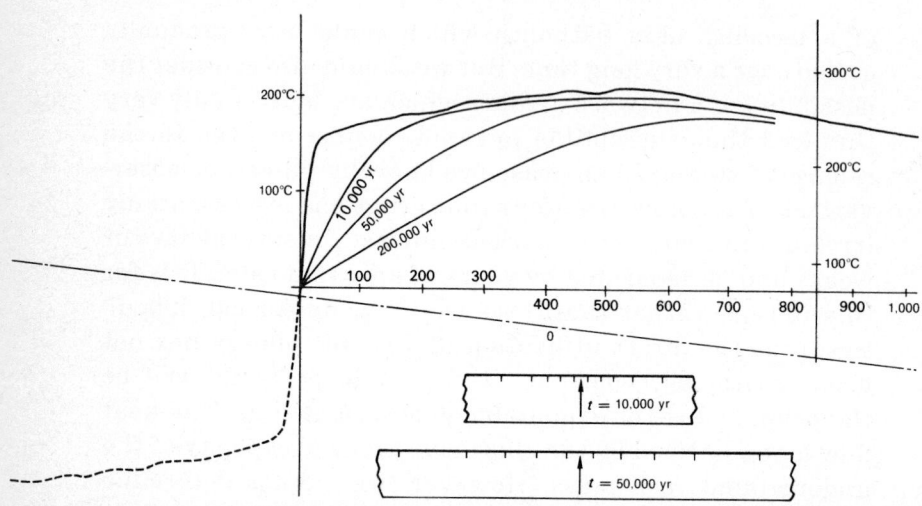

FIGURE 4-2   Calculation of the evolution of the thermal profile of a
geothermal site with heat diffusion. The thermal profile example is taken from well 48 at Wairakei; account is taken of
the normal geothermal gradient and the surface temperature. The profile was completed symmetrically. To calculate
the temperature at the end of a given time, the average of 10
values is taken on both sides of the position studied, at distances which increase as the square root of the time. Representation of two of the rules for this calculation. Observe
that at the end of 10,000 yr the useful heat which could be
extracted, (i.e., below 200 m), will be practically unchanged.

mal gradient maintained by conduction as a consequence of
the subterranean geothermal flux, and we must subtract it
from the actual thermal profile so as to define the thermal
anomaly associated with the geothermal site, which will
gradually disappear through conduction.

We must impose the condition that the temperature at
the ground surface remains constant and equal to our conventional 0°. To meet this condition, it suffices to extend the
thermal profile symmetrically, but with opposite sign,
above the earth's surface, thus assuring the constancy of
the temperature at the surface.

Now, we apply the formula which describes heat diffusion in a region, initially at a temperature $T(z)$, and which

gives the temperature after a time t as a function of z_0:

$$\theta = \frac{1}{\sqrt{4\alpha\pi t}} \int_{-\infty}^{+\infty} T(z)e^{-(z-z_0)^2/4\alpha t}\, dz \qquad (4\text{-}1)$$

From this formula we can calculate the value of the gradient at any point or at the surface:

$$\left(\frac{\partial\theta}{\partial z}\right)_0 = \frac{1}{\sqrt{4\alpha\pi t}} \int_{-\infty}^{+\infty} \frac{T(z)e^{-(z-z_0)^2/4\alpha t}}{4\,\alpha t}\, dz \qquad (4\text{-}2)$$

The calculation of an actual thermal profile, $\theta = T(z)$, must be done numerically, which means that we must replace the integral by a sum over a limited number of terms. Let us therefore take $(z - z_0)/\sqrt{4\alpha t} = s$; $I_1(s) = 1/\sqrt{\pi} \int_{-\infty}^{s} e^{-s^2}\, ds$; $I_2(s) = \int_{-\infty}^{s} e^{-s^2} s\, ds$ where $I_1(\infty) = 1$ and $I_2(\infty) = 0$. Then, formulas (4-1) and (4-2) can be written

$$\theta = \int T(z)\, dI_1 \qquad \text{and} \qquad \frac{\partial\theta}{\partial z} = \frac{1}{\sqrt{4\alpha\pi t}} \int T(z)\, dI_2$$

We will replace dI by a finite segment equal to 0.1 and will take 10 values of s at the interior of the corresponding intervals, $S_1, S_2, \ldots, S_{10}$. It will suffice to take the average of the 10 values of T, for the values of Z corresponding to $S_1, S_2, \ldots, S_{10}$, z_0 and t being chosen. For the gradient, the same average will be divided by $\sqrt{4\alpha\pi t}$.

For $N = 10$, we will take $S_n = \pm 0.088, 0.273, 0.476, 0.732$, and 1.163. For the gradient, $S_n = \pm 0.324, 0.597, 0.832, 1.097$, and 1.52.

In practice, we would graphically trace the thermal profile at the initial time (subtracting the temperature corresponding to the normal gradient) and complete this profile by its symmetry beyond the earth's surface. To determine what the profile would be at the time t, we would carry on a rule the values of $z - z_0$ corresponding to the S_n, and by centering the rule on a value of z_0, we would read the 10 values of T, whose average would give the temperature of z_0 at t.

If it is especially sought to calculate the evolution of the temperature at a point and the gradient at the surface, we can trace the thermal profile as a function of ln (z), and in the same manner the rule with the values of ln (z) corresponding to S_n, such that it would suffice to move the rule along a scale of $\frac{1}{2}$ ln t.

Figure 4-2 indicates the results of a temperature evolution calculation for the profile of shaft 48 at Wairakei, New Zealand. The evolution of the gradient at the surface has been directly calculated with the results in Table 4-1.

Table 4-1
Evolution of temperature gradient at the surface

Present	100	200	500	1,000	2,000	4,000	10,000 yr
5	1.5	1.09	0.727	0.54	0.379	0.228	0.148°C/m

From these results, we can infer that part of the rising steam has heated the surface until a very recent date—in fact, this heating is still going on—but part of the energy of the geothermal bed will hardly decrease from the value that exists today in 10,000 or 50,000 yr. The heated rock can conserve its temperature for a very long time at a certain depth, even when the surface gradient has decreased to a point where all surface manifestations have seemingly disappeared.

For a geothermal steam bed, like at Larderello, things are a bit more complicated. We can assume that under the impermeable cover the steam is at an apparently uniform temperature and pressure (for Larderello, 250°C at 40.5 bars). The surrounding groundwater can thus be in pressure equilibrium with the steam at a certain depth. For example, if the geothermal bed cools by 1°C, the saturated vapor pressure decreases by 0.685 bar, then the equilibrium level between the water and the vapor will rise about 6.85 m. The mass of vapor which is present diminished partly because of the reduction of occupied volume and partly because of the pressure drop. The latent heat corresponding to the condensation of the difference will thus be liberated. However, a more complete calculation shows that this effect is very small in comparison to the cooling of the rocky mass in which the geothermal site is included. Because of vapor convection, all the geothermal site which is occupied by steam will cool in a nearly uniform manner. The temperature of the entire geothermal site will follow the cooling of its roof, and that can lead to an appreciably more rapid cooling that in the preceding case.

This reasoning would apply to a geothermal steam site embedded in permeable rock which has a sensibly uniform temperature. The case of Larderello is different, since the steam there occupies fissures and karstic spaces and only heats the walls to a limited distance which is still not well known. The results indicated in Chap. 3 lead us to deduce a recent age for the geothermal steam site. Otherwise all the rock between the fissures would be at an approximately uniform temperature.

4-8 Rock alterations and water geochemistry

It is well known that hot water reacts with the rocks it traverses and acquires by this means chemical compositions that are different from cooler waters of the same region (to which is attributed the therapeutic properties of thermal mineral waters).

A fortiori, very hot water reacts strongly with rocks that it traverses. It is striking to see outcrops of rock in New Zealand which have been leached by thermal water and have become porous and profoundly altered to a silica-aluminous and ferric skeleton while being leached of all soluble elements such as the alkalis. Such altered rocks should draw attention, and it is surprising that they have not been pointed out more often.

On the other hand, at the time of the extraction of mineralized hot waters, especially if this operation involves vaporization, it is bothersome that drill holes sometimes become rapidly obstructed by deposits. It has been supposed that analogous deposits could have been formed in the natural terrain and could have obstructed fissures by which steam escapes to the surface. This would create an impermeable cover which would assure the conservation of the geothermal site. This process, called *self-sealing*, ought to occur at the time of vaporization of mineralized water to make the roof watertight; its reality has never been clearly established.

The chemical composition of water can furnish valuable indications of its origin, and compositional studies are indispensable in prospecting for a geothermal field. The handiest indication of the maximum temperature reached by water is furnished by its silica content, since silica does not immediately precipitate on cooling.[1] The solubility of quartz and amorphous silica increase rapidly with temperature up to around 300°C if the pressure is that of the vapor phase (much more rapidly at higher temperatures under higher pressures). We suppose that at great depths and at such temperatures, quartz is present everywhere and that the water is in equilibrium with quartz as a solid phase. At the time of its ascent to the surface, decompression of the water results in its partial vaporization which increases the silica concentration by an easily calculated amount (it matters very little whether the steam produced remains constantly in equilibrium with the water, which corresponds to an expansion at constant entropy, or is involved in a throttling process, which is a cooling at constant enthalpy). Near the surface, the water will be in contact with amorphous silica, but precipitation is sufficiently delayed so that the total silica in the water furnishes a precise indication of the initial temperature. If the initial temperature was less than 210°C, however, the silica content in the water after expansion remains less than the saturation limit for amorphous silica, and so the water can dissolve the silica in the conduit, leading to an overestimate of the temperature. Figure 4-3 shows the result of such a calculation.

Other proposed indicators which are sometimes used have a much less general application. The ratio K/Na has been proposed; it increases with the final temperature reached by the water, but it depends strongly on the local

[1] R. O. Fournier and J. J. Rowe, Estimation of Underground Temperature from the Silica Content of Water from Test Springs and Steam Wells," *Am. J. Sci.*, vol. 264, pp. 685–697, November 1966.

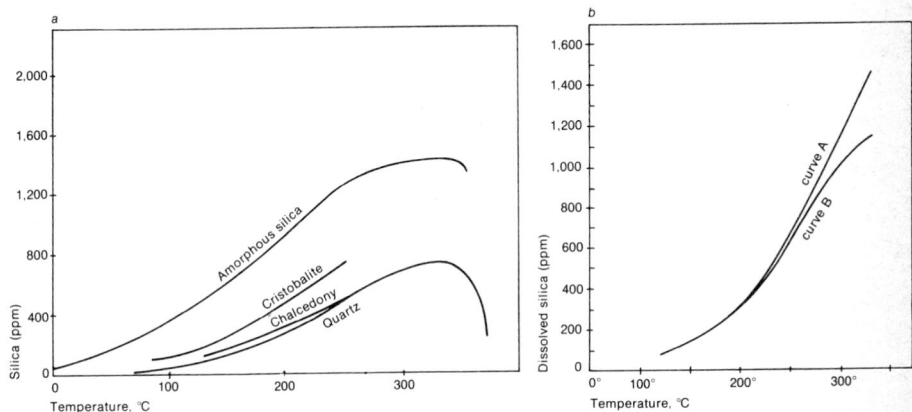

FIGURE 4-3   Relationship between the maximum temperature achieved
and the silica content. (*a*) Solubility of different silica varie-
ties as a function of the temperature; (*b*) temperature of
water at the time it was in equilibrium with quartz as a function of the measured silica concentration (B, adiabatic cooling,
i.e., a steady-state system; A, cooling with irreversible steam
expansion). (*From Fournier and Rowe, Estimation of Underground Temperature from the Silica Content of Water from
Test Springs and Steam Wells, Am. J. Sci., vol. 264, pp. 685–697,
November 1966.*)

petrography. It is an index the significance of which must be
established in each region.

When there is vaporization of water containing both
dissolved salts and gas, they are unequally distributed between the two phases. Thus it ought to be possible to recognize water coming from condensation of steam, and also,
depending on the gases present, it should be possible to tell
if a vapor is of a very deep origin or results from evaporation relatively near the surface.

The geochemical indicators require very careful interpretation which cannot be put into general rules. Only
careful comparison of the chemical composition of different
waters in a region, taking into account the petrographic nature of the rocks present, can lead to valid conclusions.

utilization of geothermal energy

The Romans used hot springs to heat their baths, and such uses are found everywhere (aside from possible therapeutic applications which have considerable economic importance). Certain Parisian swimming pools are heated directly by wells driven into aquifers of Albian greensands with a temperature of 35°C. Since the Middle Ages, localities like Chaudes Aygues, Dax, and Ax-les-Thermes in France have distributed water at 80°C for domestic and heating purposes.

Many such direct applications have developed. The entire domestic heating in Reykjavik (which is likely to be used a large part of the year) and the heating of greenhouses use geothermal waters. In New Zealand, a paper factory as well as many domestic applications (hotels, breeding farms, etc.) can also be cited.

All these applications depend on local needs. The problem of the utilization of geothermal energy does not take on all its significance until it is placed in the framework of an energy market made possible by means of electrical energy transport. Thus, we will examine the use of geothermal energy primarily to produce electrical power. We must also mention that when there exist needs for heat at a low temperature, it is more economical to omit the electrical intermediary.

All electrical or mechanical energy production from heat rests on the first two laws of thermodynamics. On the one hand, there is an equivalence between the total heat absorbed and the energy according to the relationship $4.18 \text{ J} = 1 \text{ cal}$. On the other hand, it is not possible to produce energy by the consumption of heat at a uniform temperature (this heat could be obtained in practically unlimited amounts from a source such as a river, the ocean, the atmosphere, etc.). The engine which produces mechanical energy must take heat from a warm source and give back a lesser quantity of heat to a cold source in order to transform the difference between these two quantities of heat into energy. The absolute temperatures T_1 and T_2 of these sources (in practice $273 + \theta°C$) are defined in such a way that $Q_2/T_2 \geq Q_1/T_1$, Q_1 being the quantity of heat taken from the hot source T_1. In other words, the entropy, $dS = Q/T$, can only increase. The energy produced is equivalent to the difference $Q_1 - Q_2$. In general, it is the hot source which limits the production of energy and is taken as a reference for the overall efficiency. However, the cold source does not play any less important role in the economics of energy extraction, and it is useful to have next to the power plant, as at Wairakei, New Zealand, a large river (Waikato) with permanently cold water. Otherwise one may have to use the atmosphere to cool water at the base of towers where upward movement of heated air assures a natural draft. Apart from the cost of such an installation, in some climates it does not produce a very low temperature.

Related to the heat taken from the hot source, the efficiency is

$$\frac{Q_1 - Q_2}{Q_1} < \frac{T_1 - T_2}{T_1}$$

The ratio on the right-hand side of the inequality is the thermodynamic efficiency, which constitutes an upper limit. In fact, the practical efficiency is much lower than that theoretical limit, but the discrepancy depends on technological considerations, and we can always try to reduce the difference.

What are the theoretical possibilities for geothermal energy production? Let us consider a surface of 1 km² where the regional geothermal flux is greater than the average world flux in a ratio m. This flux, $m \times 1.2\,\mu cal/cm^2 \cdot s$, is equivalent to $1.2\,m \times 10^4 \times 4.18$ W, which is $50\,m$ kW/km².

But in order that this flux be partly transformed into energy, we must tap the heat in subterranean regions which is at higher temperatures than at the surface, the heat of which represents the cold source. If the thermal gradient is m°C/30 m, the temperature difference at the surface will be $(m \times h)/30$ and the resulting thermodynamic efficiency is $(mh/30)/(283 + mh/30)$.

A hot spring represents a natural device which brings to the surface, with little temperature loss, part of the subterranean flux. There have been several developments based on the energy utilization of heat transported by a hot spring. At Kiabukwa in Katanga, there is a spring of 40 liters/s at 91°C which has been equipped to produce 140 kW.[1] But such an extraction, which could continue indefinitely, is highly unusual. As a general rule, a close inspection of different developments shows that the heat contained in a certain volume of rock is used up. This heat is transported to the surface by water circulation in either the liquid or the vapor state. Such a geothermal site is thus likely to deplete itself, the more quickly the larger the flow, unless the subterranean phenomena which gave rise to the geothermal site renew themselves. But the way in which these original phenomena are distributed in time remains obscure.

It is easy to calculate that a cubic kilometer of rock cooling from 250 to 200°C, liberates 25×10^{15} cal, which is equivalent to 10^{17} J. By taking an average efficiency of (225°C − 30°C)/(273°C + 225°C) = 38 percent, we could theoretically produce 12,000 kW for 100 yr.

Although this is very theoretical, we can undertake a more precise calculation. Let us take 27°C, which is 300°K, for the surface temperature and suppose a gradient of $30m$°C/km, which gives $T = 300$°C $+ 30\ mh$°C. If we remove some of the heat contained in 1 km³ thus decreasing the temperature from T to $T - dT$, it will provide $0.5 \times 10^{15}\ dT$ cal, with a theoretical efficiency $(T - T_0)/T$, which can produce a total energy of $2.1 \times 10^{15}\ \int_{T_0}^{T} [(T - T_0)/T]\ dT = 2.1 \times 10^{15}$

[1]A. Rollet, "La Centrale Géothermique de Kiabukwa," Report of the Congrés Scientifique, fiftieth anniversary of the Special Committee for Katanga, communication 96, Elisabethville, 1950.

$[T - T_0 - T_0 \ln (T/T_0)]$, or approximately $2.1 \times 10^{15} (T - T_0)^2/$
$2T_0 = 1.05 \times 10^{15} (30mh)^2/300 = 10^5 \ m^2h^2$ kW $\cdot$ yr.

If we extracted all the ground heat to a depth of H km, we could theoretically produce $(10^5/3)m^2H^3$ kW $\cdot$ yr/km². Presently only thermal sites with high thermal gradients are used; m can attain a value of 10, but only over small areas and to a shallow depth H. However, if this formula is extended to all the land area above sea level, which is 142 million km², we would find a total of $4.8 \times 10^{12} \ H^3$ kW $\cdot$ yr. Although in extracting this heat one would encounter considerable technological problems, we should remember this figure because it sets practically no theoretical limit to the possibilities of geothermal energy production. Now it remains to discover how to produce this energy and at what cost.

Practically, we are able to distinguish three cases in the operations technology: first, the geothermal site which only furnishes steam; second, the site which supplies very hot water which can be partially vaporized by a pressure drop, and this subsidiary steam used for energy production; and finally, the site in which the heat directly contained in the water is used without vaporization. We will examine these cases in succession.

5-1 Operating techniques for a steam site

The classic example of this case is the steam site at Larderello[1] in Tuscany, where the first energy was produced at the start of the century and where the production of energy has now reached 200,000 kW. The geothermal site at the Geysers, in California, also uses a steam field which extends over 25 km². This site reached 396,000 kW (October 1973) and has continued to increase. It is expected to reach

[1]A. Mazzoni, "I Soffioni boraciferi Toscani e gli impianti della Larderello," Bologne, 1951. Also see the reports given at the United Nations on new energy sources, Rome, 1961, and at the Symposium of the United Nations on the Development and Utilization of Geothermal Resources, Pisa, 1970.

600,000 kW. There is also a steam field which is utilized at Matsukawa, Japan.

The sinking of commercial wells involves precautions which we will find in all the cases. Before reaching the roof of the bed, it is necessary to have cemented in a casing that can withstand the pressure exerted when the main valve is closed without the risk of vapor infiltrating into the ground. After the hole has been drilled using refrigerated mud, with normal precautions to control possible eruptions (fast-acting valves called *blowout preventers*), a protective column can be installed with a casing or the productive zone can be left open. At Larderello, a drill hole was generally stopped as soon as a steam-producing fissure was reached. In principle though, with a sufficiently cold mud, drilling could continue to obtain higher production capacity.

Once the hole is opened and emptied, the steam escapes into the atmosphere. Then, once the walls of the shaft are at temperature equilibrium, the steam expansion becomes adiabatic. It is known that a saturated vapor in an adiabatic expansion to atmospheric pressure becomes humid. That is, the steam contains about 25 percent (by mass) of water in suspension. The steam will appear as a white plume; at the end of several very spectacular days of this free release into the air, the plume becomes transparent and only condenses higher up by mixing with the colder air. Under atmospheric pressure, the steam drys up. We can interpret this change by supposing that the expansion from the geothermal pool no longer takes place in the drill holes but a little bit upstream, and the steam cooled by expansion is reheated on contact with hot rock remaining at 250°C (in the case of Larderello). Of course the propagation of the pressure drop due to the release of steam in a fissure network whose form we do not know cannot be described accurately. We only know that the steam with a lowered pressure takes heat from the surrounding terrain. At that time, the drill hole is connected to the central power station by insulated piping of a large diameter (massive flows can be hundreds of tons an hour, and a ton of steam occupies 279 m³ at a pressure of 7 bars). The pipes are arranged so that they can freely dilate, but there is no worry about water condensation since the steam is superheated.

Formerly, to avoid corrosion of the turbine, an exchanger was used to vaporize pure water while the natural steam condensed. Naturally, there was a loss of temperature and pressure in this operation. There was then no problem in using a condenser where the final condensation of the steam occurred at a low temperature and pressure.

Then, it was decided to construct turbine vanes out of steel resistant to the corrosion of the natural steam. But the gas contained in natural steam (5 percent at Larderello, 1 to 2 percent at the Geysers, where this proportion is dropping in the course of the operation to 0.5 percent) accumulates in the condenser increasing the pressure, which has to be lowered by pumps. In spite of that, the pressure in the condenser is never as low as with a boiler using recycled water.

A third solution, more primitive and permitting only a marginal efficiency (that comes from taking 100°C as the temperature of the cold source), consists in letting the turbine exhaust go directly into the atmosphere. This method is sometimes used for a short while by a field unit to utilize the production of a drill hole until it is connected with the power station.

5-2 Directives for geothermal operations

A number of drill holes are connected by pipelines to the entrance of the turbine. There can be two or more separate networks connected to different turbines which can work at different pressures or a joint collection allowing several turbines to be fed. But in any case, the turbines are calibrated for a certain entrance pressure, which means a slightly higher pressure at the wellheads. Regulation of flow, as a function of the electrical power to be provided, is controlled by regulators at the turbine entrance which maintain the steam flow at a certain pressure. Sometimes, in the course of the operation, that entrance pressure is appreciably different from the design pressure. In the power station's operations, it is necessary to choose between maximum energy production, which corresponds to a strong flow with a relatively low upstream pressure, or a high efficiency in kilowatts per kilogram of steam per hour, which requires a high upstream pressure with a lesser flow.

Even if it was demonstrated that the total quantity of heat which can be extracted from the ground, and thus the total steam tonnage which the site would be able to produce, is limited, producers will rarely sacrifice the possibilities of a high production capacity in the immediate future for the benefits of prolonging the site's reserves, since its limitation will not appear until after a very long waiting period. In the establishment of the power-station plan, the upstream pressure at the turbines should be fixed (or possibly the pressures if there are two groups of wells, stronger and weaker producers, feeding different turbines). This is a delicate choice which requires a thorough knowledge of the production capacities of the wells, at a time when development wells are hardly completed or are still to be drilled. In practice though, it is often found that the power plant is working at a pressure lower than designed. Should we thus conclude that the production pressure was chosen too high or the plant overequipped? The constant concern of the producer is to increase the production capacity by sinking new wells extending the surface area of the exploited field.

These new wells are especially necessary as experience has shown that well production decreases with time (for example, to one-half in 2 or 3 yr in Larderello, although exact information has not been published). One of the explanations of this decline holds that the ground heat used to superheat the steam is depleted further and further upstream from the fissures by which the steam reaches the drill holes.

5-3 Geothermal exploitation involving water vaporization

An example of this type of exploitation is that at Wairakei, New Zealand.[1] This site is equipped to produce 175,000 kW. The proximity of the Waikato River, which originates from Lake Taupo with a large flow, was used to advantage by installing the power plant on its banks, permitting very efficient operation of the condensers. However, this extends the piping a bit and inhibits the development of the extension of the geothermal field that was discovered on the opposite side of the site.

A well penetrating into a bed where the water temperature is at the boiling point under the appropriate pressure does not risk eruption whenever the mud is cooled. Yet, it may be necessary to maintain the cooling mud circulation while the drilling operations are halted. In New Zealand, they did not limit themselves to cementing only the first casing in the ground, but immediately set a truly massive mooring, which is perhaps dispensable. A casing is set slotted at a suitable level even for exploratory drill holes.

After 3 months of waiting, during which the main valve is closed, the well comes into equilibrium with the surrounding ground. The water at the bottom of the well is in equilibrium with the ground, but gas can accumulate and mix with the steam; this gas occupies a certain height in the column so that the valve pressure has no particular significance.

After the temperature profile is measured, production tests can proceed. If steam accumulates, it suffices to open the main valve; the pressure falls in the well, and the entire column of water begins boiling, the steam jet driving out the water (and possibly all the rocky debris which has accumulated in the well). If the eruption does not begin spontaneously, it is necessary to raise part of the well's water

[1]G. W. Grindley, The Geology, Structure, and Exploitation of the Wairakei Geothermal Field, Taupo, New Zealand, *N. Z. Geol. Surv.*, Bull. No. 75, p. 131 (geological map and cross sections), 1965. See also the numerous publications of the Department of Scientific and Industrial Research of New Zealand in the form of internal reports, or the *New Zealand Journal of Geology and Geophysics.*

column to initiate the eruption. This can be done by a process called *air lift*. Essentially, this involves injecting compressed air at a certain depth in the well.

In the course of the eruption, water flows from the ground toward the well, and either in the fissures or later when passing the slotted casing, part of this water is vaporized because of the pressure drop; the temperature then decreases. We can easily calculate the proportion, by weight, of water transformed into steam. Let θ_1 be the initial temperature, θ_2 the boiling point of water at the pressure in the well, and L the latent heat of vaporization at that temperature. The vapor proportion x is such that $x \cdot L = \theta_1 - \theta_2$. Although x is much less than unity (25 percent at most), the vapor volume is large with respect to the water volume, and the water is reduced to a suspension in the steam jet which can reach a high velocity. This suspension could be formed in a well open to the atmosphere as well as in a hole with a back pressure maintained at the well's main valve.

When a well furnishes superheated steam, as at Larderello, the flow can be measured by one of the methods applicable to gases, such as the pressure difference on the two sides of a calibrated orifice, or the pressure measured upstream from a conical fitting which leads into free air and which functions like a venturi. However, these methods are not applicable to a humid vapor. In New Zealand the production is measured while a back pressure is maintained by a calibrated orifice, or *duse*. The pressure is measured upstream from this orifice. The steam content and the flow of the mixture are measured downstream after a new expansion which increases the amount of steam. To determine this amount (see Fig. 5-1), part of the steam jet is collected and is condensed in a receptacle containing cold water. The increase of mass and of temperature give the proportions of water and steam. The total flow is estimated by the ratio of the total cross section of the jet to the sampled cross section. Care must be taken to move the sampling device over the entire area of the jet during testing. With different orifices,

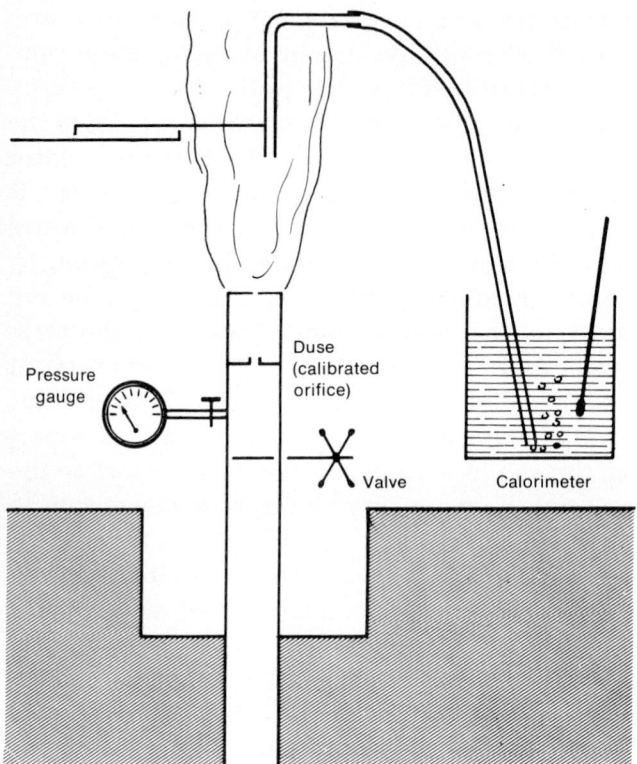

Pressure gauge

Duse (calibrated orifice)

Valve

Calorimeter

FIGURE 5-1   Principle of the production measurement of a geothermal well. Different outputs, characterized by their pressures at the wellhead, are obtained with different *duse* (calibrated orifices). After expansion, a fraction of the flow is removed from the steam jet (a calibrated cross section being displaced for a systematic sampling), and the product is then condensed in a calorimeter and the heating and mass increase are measured.

one can determine the well's production as a function of the pressure at the main valve, and thus is able to choose the range of turbine operation. It is necessary to repeat these measurements periodically.

To utilize the well, we must first separate the water and the steam. This is done in a large vertical tank (Fig. 5-2) which the steam jet enters tangentially. The steam is extracted from the upper part of the tank and the water ac-

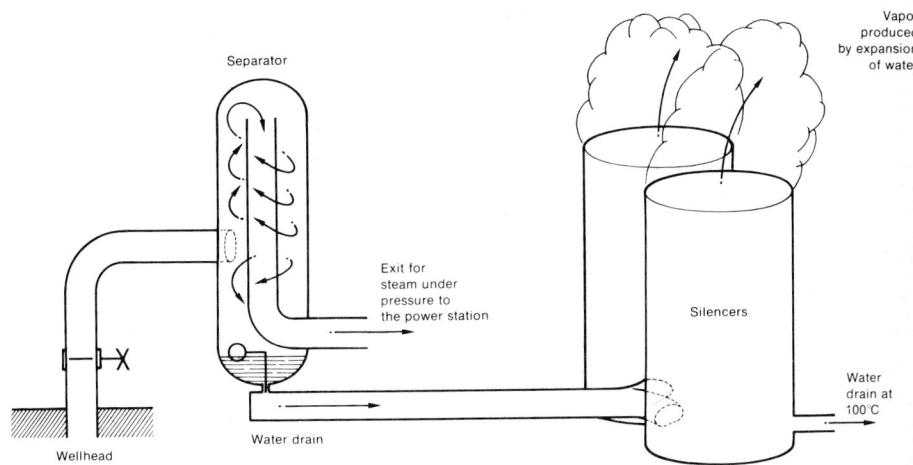

FIGURE 5-2 Diagram of the wellhead equipment used at Wairakei. A separator and a muffler are found in which water is successively separated under pressure and expanded with partial vaporization up to atmospheric pressure.

cumulates at the base. The temperature of the steam is a function of the pressure at which the vapor is used, but in any case it is much higher than 100°C. On further expansion, the water that has been collected partially vaporizes and the steam thus produced could be used by a second collection network at a lower pressure. As there are two collecting networks, one at high pressure and one at low, this seems easy. If this is not done, it is simply because it appears more economical to increase production by sinking new wells. It is evident that in this comparison no consideration has been given to the total energy which will be produced during the life of the geothermal field. The heat contained in the geothermal site will dissipate at a higher rate in the course of the operation than if thermal recovery had been allowed.

It is even theoretically possible to use the steam produced by the expansion of water to atmospheric pressure (and even use the heat of water below 100°C), if the condensers can be arranged downstream from the turbines. This will be seen later.

But this recovery is not done at Wairakei, where the water is flashed at a temperature greater than 100°C (up to 160°C, at which temperature steam production by expansion reaches 15 percent). It is the noise produced by this expansion which is the most irritating aspect of a geothermal energy operation, and to reduce it water is expanded in *silencers* formed by two vertical cylinders coupled at the base where water enters. The escaping steam condenses in the cold air producing white steam plumes (actually, mist).

5-4 The evolution of a geothermal site

We have assumed that at the beginning of the exploitation, the geothermal water is vaporized on reaching the well or while traversing the slots in the casing. An evolution of producing conditions is observed at Wairakei that suggests that this is not always the case. It seems that, in the long run, a pressure drop propagates upstream and leads to a partial vaporization of the groundwater. It is clear that at each point the vapor will tend to occupy the upper regions of the permeable reservoir, near the impermeable cover. But this cover cannot be totally impermeable, and the steam must infiltrate it. As a result the superficial manifestations (i.e., geysers, mud pots, etc.), which made Wairakei a tourist area long before geothermal operations began, tend to increase instead of being reduced as was anticipated. The steam proportion in the well production has increased, which implies a vaporization not only by expansion but by contact with rock having sufficient thermal inertia to supply the necessary heat for vaporization, due to a pressure drop in the bed. It is evident that this forms a very favorable opportunity for exploitation.

But it is not easy to tell to what distance from the wells this vaporization could propagate in the surrounding ground. If we imagine the case of a permeable matrix in which vaporization takes place, we realize that (taking account of the volume flows of the water and steam, for similar mass flows and actual viscosities) the pressure gradients must be much stronger in the region of the steam than in the region occupied by the water. Hence we might expect the matrix (at Wairakei, it is tuffs and volcanic cinders that are rather friable) to burst in the area of vaporization. It is likely that this occurs in uncased wells, where the jet of steam and water is often loaded with rocky debris.

It is thus likely, especially taking into account the large production differences between neighboring drill holes, that fissure permeability plays a larger part than matrix permeability. We can conceive that, in a fissure with easy access to a well, water is ejected by vaporization and this process results in the creation of a system of communicating fissures in the well walls filled by steam until ultimately water is replaced by steam in the matrix itself.

Ideal operating conditions would be realized if, in the layer which constitutes the geothermal reservoir and whose upper regions are perhaps filled with steam, at least near the well, only the upper region of the layer interacted with the drill hole. Then vaporization would be produced in the layer itself, and at the limit, we could extract only the steam with all the advantages which have been previously indicated. This could resolve the problem of what to do with the residual hot water. Also, this would limit a great deal the problem of resupplying water to the geothermal site. The vaporization of the water contained in a rock having a porosity of 10 percent effectively lowers its temperature 80 or 90°C, which closely corresponds to all the heat which can be usefully extracted.

If we recall the technological progress which has been achieved in the development of oil wells, we can hope that the same will develop in the future exploitation of geothermal sites and that we will thus succeed in extracting only the steam contained in a bed initially occupied by water. This is definitely in the future, however, and it will be indispensable to understand, much better than today, how vaporization propagates in the matrix as well as in the fissure network.

5-5 The utilization of steam

The steam gathered in separators is sent to turbines by a system of pipes. But, as this is a saturated vapor, it constantly condenses and the water flows to the bottom of the pipeline, which must be automatically purged at all low points. The simultaneous flow of both water and steam in the same pipe can be unstable and should be avoided. The use of steam at the power station equipped with turbines and condensers presents no new insights.

Observe that in both the modes of operation which have been described, a certain number of wells are linked in parallel to the same collector network. It should be understood that there can be several different supply networks utilized at different pressures, pressures which are defined downstream by the functioning of the turbines. Thus, we do not know which wells supply the steam utilized, which in the long run could be very inconvenient (a well could effectively stop producing, without notice). This makes it necessary to take periodic measurements, in practice monthly, of the conditions under which each well is producing steam.

In the design of a commercial energy project, we have pointed out that the first problem is the choice of the operating pressure. Another problem concerns the number and position of the power stations. On the one hand, it would be useful to reduce the length of the steam pipeline, which is costly and the site of heat losses. But also, it is desirable not to disperse the power installations too much in order to use large steam engines, which are more economical than small ones. These conditions lead to a compromise, with some dispersal of the power stations. At the Geysers (California), where the first power station reached 14 and then 28 MW, they plan to reach 600 MW at the end of 1975 (without this being an upper limit), with power stations including two groups of 55 MW each. However, local conditions, particularly the availability of cooling water for the condensers, make a detailed study necessary for each case, since general rules cannot be formulated.

5-6 Energy utilization without steam production

There is no fixed limit between the application of this method and the preceding method. Theoretically, we can vaporize water at less than 100°C in a partial vacuum created by an efficiently cooled condenser downstream from the turbine. This is how the small installation at Kiabukwa worked. As is done in Reykjavik, water can also be used at 140°C for urban heating without allowing it to vaporize, by maintaining the water in a pipeline system which is under pressure. However, for this, pumps must be immersed deeply in the drill hole so that the pressure of the extracted water is always at least of the order of 3 bars on the surface. In Reykjavik, this pressurized hot water is used in exchangers for domestic water heating and for heating greenhouses.

If it is necessary to produce energy, steam turbines operating at very low pressures should be avoided because dissolved gases accumulate in the condenser, increasing the pressure, which then must be lowered with pumps. Instead, closed-circuit thermal engines operating between two exchangers which respectively constitute the boiler and the condenser could be used, with a turbine in the steam path and an injection pump to supply the condensed liquid to the boiler.

This circuit can just as well be used with a liquid more volatile than water, and thus the vapor pressure will be higher at operating temperatures, and turbines and pipes can be smaller. In China (near Peking), ethyl chloride is used. Butane, propane, or even Freon or ammonia could also be used. It can be observed that the technology necessary to make such circuits pressure-tight, while extracting the mechanical or electrical energy produced, is the same as for refrigerators, and everyone knows that leaks are no longer a problem there.

However, the theoretical efficiency remains limited by the second law of thermodynamics, whatever the fluid employed. The cost of an exchanger, overall, should be proportional to its exchange capacity, and it becomes much more costly as we try to reduce the difference between the secondary fluid exit temperature and the primary fluid entrance temperature. The establishment of a project necessitates an optimum choice between the installation cost and the efficiency. There do not yet seem to be any rules which allow a calculation of the cost of an installation as a function of its output power, but undoubtedly such installations are possible (witness the Chinese experience).

Anytime there exists a need for heat at a low temperature, whether for domestic heating or for certain industries such as the desalination of water or brine (whether or not this is the water in the geothermal bed), there is great interest in utilizing the heat of the water extracted from the ground directly without the intermediary of electrical energy. For heating, electrical energy is generally employed through the joule effect, which means the transformation of electrical energy to heat through a resistance. Thus, we do not retrieve all the energy spent to produce the electrical energy, which lowers the effective efficiency to less than the theoretical limit of the thermodynamic efficiency, which is very marginal for waters at relatively low temperatures.

It would not make any difference if we used a heat pump for the electrical heating, which is equivalent to a refrigerating engine and which utilizes the heat which a refrigerator dissipates in the ambient air. It would be forced to cool some external source (the atmosphere, for example). We pay little attention to the cost of cooling by our refrigerator since it is the only way this cooling can be done. But for the calories produced by a heat pump, the comparison of cost with the cost for producing heat by the joule effect is striking. Equipment for heating by the joule effect, largely composed of resistances, is remarkably simple and economical. The equipment cost for a heat pump is much higher and even though the practical efficiency is higher than for the resistance heater, the savings in running costs (consumption of electricity) are insufficient to justify the initial cost of the equipment. In fact, no heat-pump design has ever gained undisputed advantage in this economic comparison.

Thus, the problem of the use of a heat pump has no relationship to the utilization of geothermal energy, especially at low temperatures. It only concerns the best way to utilize electrical energy.

We will not enter into the details of the utilization of geothermal waters for heating at low temperatures except to point out that the degree of water mineralization often requires the use of a heat exchanger, and that disposal of the pollution by mineralized water can be difficult.

The utilization of water from natural thermal sources or from artificial drill holes poses no particular problems. It is applicable on a large scale in a country like Hungary, where there is no hesitation in exploiting deep aquifers with wells 1,500 or 2,000 m in depth.

5-7 Water injection

In many other cases than those just mentioned, the supply conditions for subterranean water layers, which are often captive, are not such that we can expect steady large flows for profitable geothermal energy extraction. This stimulates the idea of artificially supplying subterranean water layers by injecting, in suitably placed wells, cooled mineral water, which we have already seen is a nuisance after the heat has been extracted.

This artificial supply can take very different forms. The project undertaken in the Imperial Valley, in the southern part of California (which is partially associated with the hypermineralized site of the Salton Sea), will utilize an apparently massively permeable basin. It has been proposed to inject water in the cooler marginal areas of the basin. This injection would simply add to the actual water supply outside the convection cells.

But if we consider an aquifer at a relatively uniform temperature, it could be advantageous to inject cooler waters directly into the hot aquifer. The model of such an injection was done at Melun (near Paris) and assures a supply of hot water and heating to nearly all of a large set of apartment buildings (a supplementary use of heating fuel is necessary a few weeks each year). From petroleum prospecting, it is known that in the Bajocian, at a depth of 1,800 m, there is a very permeable aquifer at a temperature of 70°C. This bed has been reached by two separate wells drilled in approximately the same spot with deviations in opposite directions. One of these serves to extract hot water, and the other well injects this same water after cooling through heat exchangers. The thermal-siphon effect, which may be expected, is insufficient during the winter to assure the necessary flows, and pumps are used to inject the cold water.

5-8 The future of an operation with reinjection of water

If we imagine the long-range future of such an operation—and all operations which could be achieved under the same conditions, since the geothermal site at Melun is known to extend over thousands of square kilometers—we must be concerned about the risk of the injected cold water cooling a progressively increasing area of the geothermal bed until finally the temperature of the extracted water is lowered. The calculation of the predicted temperature evolution of such waters can be accomplished in the following manner.

We will neglect the dependence of the viscosity on temperature and thus will consider it as a constant. This will lead us to overestimate the cooling of the extracted water since the viscosity variations tend to slow down the movement of cooler waters and thus increase the proportion of warmer waters in the supply networks of the wells. The hydraulic system will thus be steady, and it is easy to determine the system (see Fig. 5-3) in the case of any number of extraction wells P_i and injection wells Q_i, each employed at a constant flow (the sum of the flows not necessarily being equal). We can take into account a steady circulation in the aquifer, which is equivalent to placing two wells, P and Q, at plus and minus infinity. Observe that for a single well, the

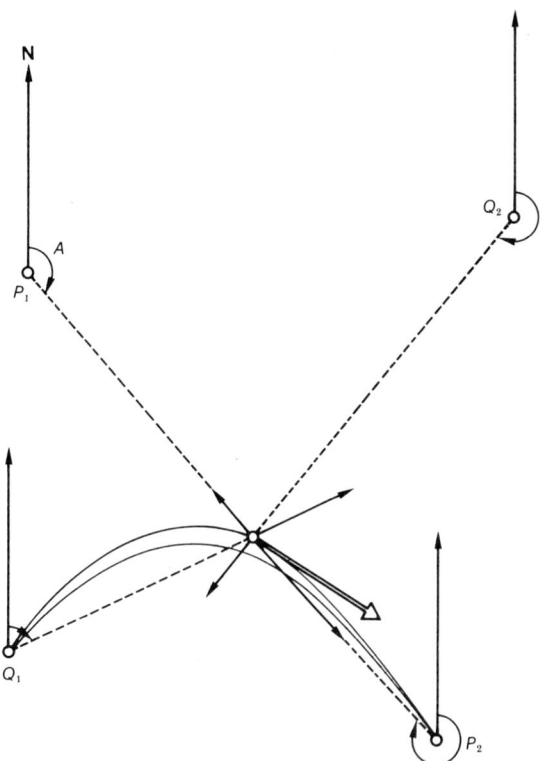

FIGURE 5-3   Principle of the calculation of current filaments circulating in a permeable layer between injection and extraction wells: geometrical definition (in consequence of the weighted sum of the azimuths being a constant), and calculation of the instantaneous flow.

unitary-flow vector is radial and inversely proportional to the distance to the other wells. In addition, the flow passing between two points is proportional to the angle of the radius vectors. For any number of wells, we obtain the unitary flow at a point by adding the unitary-flow vectors for all the wells. For a flow line, a weighted sum proportional to the flows of the wells, the azimuths of the radius vectors are constant. In the case of two wells P and Q, with equal flows, this last rule shows that the flow lines are circles passing through wells P and Q. But we could also consider two wells with unequal flows and an aquifer with a steady inflow.

We will suppose that we have effectively traced the flow lines bordering on a well P and determined the width of a current filament of known flow along this line. The width will be inversely proportional to the unitary-flow vector.

We will study[1] separately the thermal evolution along each current filament as they are defined at the end of the calculation by the principle just indicated. The current filament includes, upon leaving the well Q, between the directions Φ and $\Phi + \Delta\Phi$, a flow $\Delta Q = Q \, \Delta\Phi/2\pi$. At a distance l from the origin, its width is $e \, \Delta\Phi$, and we will have to consider its surface $s \, \Delta\Phi$, s being defined by $s = \int e \, dl$. When reaching the extraction well P, s has the value S.

Before injection, a certain temperature distribution holds, constant and in equilibrium with the geothermal flux and the surface conditions. Let θ_0 be this steady-state temperature of the aquifer, which we will consider as uniform, neglecting the variation over the thickness of the layer. When the system is perturbed by the injection of cold water, the perturbation to the temperature will satisfy, outside the aquifer, the heat equations without taking into account the geothermal flux. Moreover, the perturbation must be zero at the surface, a condition which can be satisfied by the method of images, but that we will neglect here because the aquifer is at a relatively great depth.

If θ_0 is the initial uniform temperature in the layer and θ_1 the temperature of the injected water, we will let $\theta = \theta_0 + \tau(\theta_1 - \theta_0)$. Outside of the aquifer, $\tau(\theta_1 - \theta_0)$ will be the thermal perturbation calculated with respect to the initial local temperature.

In the first phase of the calculation, let us treat the walls surrounding the aquifer as insulating. Let h be the thickness of the aquifer and C its heat capacity per unit volume. The unitary flow is then given by $q = \Delta Q/\Delta\Phi \, eh =$

---

[1]This calculation is due to M. Alain Gringarten (oral communication), whom I wish to thank for allowing me to present it here.

$\Delta Q/2\pi eh$. We will assume that the temperature stays uniform in a cross-sectional area of the filament considered, and we will neglect the heat transfer by conduction in comparison with the transfer resulting from water circulation. Then the heat equation reduces to

$$C\frac{\partial\theta}{\partial t} = -q\frac{\partial\theta}{\partial l} \qquad (5\text{-}1)$$

On introducing τ and replacing $\partial\theta/\partial l = e\,\partial\theta/\partial s$, Eq. (5-1) becomes

$$\frac{\partial\tau}{\partial t} = -\frac{Q}{2\pi hC}\frac{\partial\tau}{\partial s} \qquad (5\text{-}2)$$

All solutions of the form $\tau = F(t - s \cdot 2\pi hC/Q)$ satisfy Eq. (5-2). In other words, the profile of the thermal variation caused by the injection propagates in the layer, with respect to the variable s, without change. If the injection of cold water begins at a time $t = 0$ with a uniform flow, the thermal wave will initially have an abrupt front, and theoretically it would continue to propagate with this abrupt wave front and a velocity $\partial s/\partial t = Q/2\pi hC$, or a true velocity $\partial l/\partial t = Q/2\pi hCe$.

It is very evident that with a constant width e, the thermal wave cannot progress without its profile being attenuated, especially if it was abrupt at the origin. A primary cause is the horizontal thermal conductivity of the layer, which, if the initial wave front is abrupt, will give a smoothing to the profile with time that could be calculated by means of the diffusion equation. But this diffusion occurs with respect to the length l of the filament, whatever its width, and it would not be convenient to introduce it into the equations. Perhaps a more important factor that smooths the thermal profile is the fact that the permeable water layer is actually composed of beds having very different permeabilities, where the water, and in consequence the thermal perturbation, progresses at very different velocities. We can infer that a certain amount of diffusion reestablishes the uniformity of temperature between different layers by water exchanges between fast- and slow-moving filaments, but this process only occurs over a distance which is not negligible. These two processes have the character of a diffusion. They will smooth the thermal profile while retaining its median position, and their effect will be greatly lessened as the profile becomes smoother. We can expect that the diffusion progresses as the square root of the time. But perhaps we should analyze the diffusion with the true length of the filament l and not s.

In spite of these restrictions, when we calculate the thermal perturbations at the extraction wells P_i, the approximation of an abrupt thermal wave front is useful. We will consider for each instant the proportion of supply filaments whose temperature has been perturbed, which will permit us to calculate the resulting temperature of the assemblage of filaments. The modification which needs to be added to this perturbation curve to take account of the fact that along each filament the thermal wave has a smoothed profile, and is not abrupt, will only slightly change the initial part of the curve.

As a case in point (see Fig. 5-4), we will perform this calculation for two wells, extraction and injection, with equal flows. Then the lines of flow are circles, and if $2E$ is the distance between the shafts, a simple geometric calculation gives $S = 2E^2(\sin \Phi - \Phi \cos \Phi)/\sin^3 \Phi$ (this trignometric

Table 5-1

Φ, degrees	0	10	20	30	40	50
$t(Q/4\pi E^2 Ch)$	$\frac{1}{3}$	0.3463	0.3526	0.3728	0.409	0.456
τ	0	0.055	0.11	0.16	0.22	0.27

90	100	110	120	130	140	150
1	1.46	1.924	2.94	4.96	10.75	22.9
0.5	0.55	0.61	0.66	0.72	0.77	0.83

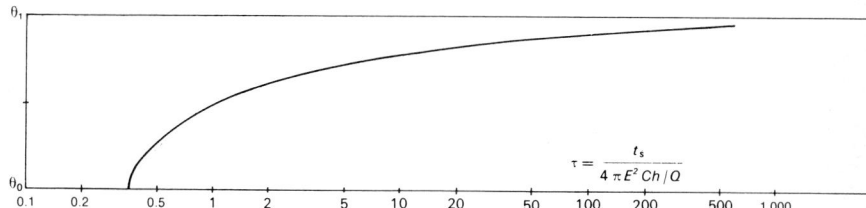

FIGURE 5-4   Calculation of the temperature at the extraction well for a pair of drill holes, supposing a permeable layer of thickness h, heat capacity C cal/cm³, and a flow Q g/cm·s. The walls are assumed insulating; the time is expressed in a reduced form as a function of the distance between the shafts, $2E$, and h, C, and Q.

expression is equal to $\frac{1}{3}$ for $\Phi = 0$). Thus, the arrival time of the thermal wave at the extraction well is

$$t = \frac{(4\pi E^2 Ch/Q)(\sin\Phi - \Phi\cos\Phi)}{\sin^3\Phi} \qquad (5\text{-}3)$$

The trignometric expression takes, as a function Φ, the following values. The corresponding values of τ have also been indicated in Table 5-1.

Let us perform again the previous calculation, but this time taking into consideration the conduction of the layers surrounding the aquifer. Hereafter, z will be the distance to the edge of the aquifer on either side. In the walls, we will ne-

60	70	80
0.483	0.639	0.882
0.33	0.38	0.44

160	170	180
74	591	∞
0.89	0.94	1

glect the horizontal conduction in order to only consider the vertical heat flux. Thus we will only have to consider the equation

$$\frac{\partial \theta}{\partial t} = \alpha \frac{\partial^2 \theta}{\partial z^2}$$

which leads to

$$\frac{\partial \tau}{\partial t} = \alpha \frac{\partial^2 \tau}{\partial z^2} \tag{5-4}$$

In the aquifer, it is necessary to take into a thermal balance the loss of heat toward the walls, $2K \, \partial \theta / \partial z$ per unit surface, which gives the equation

$$hc \frac{\partial \tau}{\partial t} + \frac{Q}{2\pi} \frac{\partial \tau}{\partial s} - 2K \frac{\partial \tau}{\partial z} = 0$$

Letting $t' = t - (2\pi hC/Q) \cdot s$, this equation becomes

$$\frac{Q}{2\pi} \frac{\partial \tau}{\partial s} = 2K \frac{\partial \tau}{\partial z} \tag{5-5}$$

Eq. (5-4) not changing form. Equation (5-5) is satisfied by a function of $z + (4\pi K/Q) \cdot s$, and $\tau = \text{erfc}\,[(z + (4\pi K/Q) \cdot s)/\sqrt{4\alpha t'}]$ follows. Returning to real time t, we find

$$\tau = \text{erfc} \left\{ \frac{[z + (4\pi K/Q \cdot s]}{[4\alpha t - (8\pi Kh/Q) \cdot s]^{1/2}} \right\} \tag{5-6}$$

We obtain the temperature value of the current filament upon its arrival at the well P by letting $z = 0$ and replacing s by its corresponding value S.

A statistical calculation for all the current filaments feeding the well P, which arrived from diverse injection wells, possibly at different temperatures or from infinity where the temperature always equals its initial value, would furnish, as a function of time, the temperature of the extracted water.

But the integration for the different filaments of that expression, where S is given by the geometry as a function of Φ, is not very convenient. The calculation is simplified if

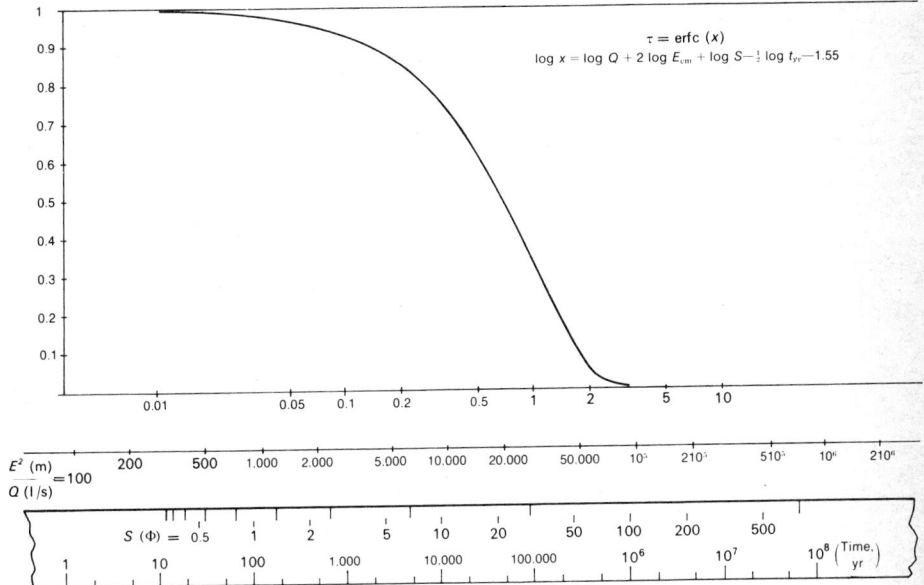

FIGURE 5-5   Calculation of the temperature evolution for the hypothetical case of a thin permeable layer with conducting walls. On the graph with abscissa of ln (E^2/Q) (m·liters/s), the form of the sliding rule having a gradation in time (years) has been calculated, the value utilized being put opposite the value of E^2/Q. For each filament characterized by $s(\Phi)$, the proportion of the temperature drop will be read on the other scale, from which a weighted average of the diferent filaments will be taken. The values of S for 10 filaments with equal flows have been indicated for a pair of wells.

the aquifer is thin (in the limit, it could reduce to a simple fissure) and if we can neglect h. By attributing all the heat capacity to the wall rocks rather than the aquifer, assuming the cross-sectional area to be uniform, we will not commit a very great error. Then from Eq. (5-6), we have $\tau = \text{erfc}\ (2\pi KS/Q\sqrt{\alpha t})$.

To do the numerical integration in $\Delta\Phi$, it will be convenient to let $\tau = \text{erfc}\ (x)$ and ln $x = \ln\ (2\pi K/Q\sqrt{\alpha}) + \ln S(\Phi) - \frac{1}{2}\ln\ t$ (see Fig. 5-5). In the preceding hypothetical case of two wells with equal flows, ln $x = \ln\ (2\pi K/\sqrt{\alpha}) + \ln (2E^2/Q) + \ln\ [(\sin\ \Phi - \Phi\ \cos\ \Phi)/\sin^3\ \Phi] - \frac{1}{2}\ln\ t$. We will con-

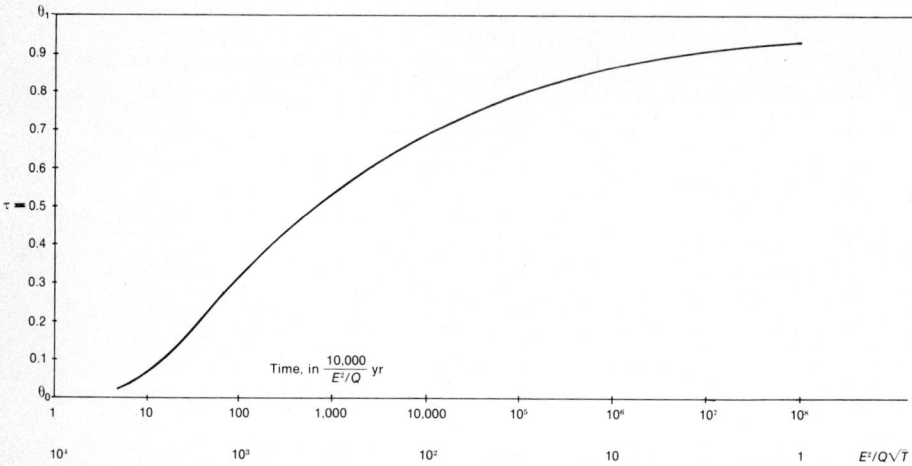

FIGURE 5-6   Curve giving the cooling behavior for a pair of wells with the hypothesis of a thin aquifer and conducting walls, E is in meters, Q in liters per second, and T is in years.

struct the curve giving τ as a function of ln x (see Fig. 5-6), and on a rule we will have the values ln X for regularly spaced values of Φ. By moving the rule according to the values t, we read diverse values of τ, from which it will suffice to take the average.

Figure 5-5 shows the rule for the case of two wells of equal flow just studied. The flux of one of the wells has been divided into 20 equal filaments, each characterized by the median behavior of the current line. For $\Phi = 9, 27, 45, 63, 81,$ 99, 117, 135, 153, and 171°, the result of the calculation is given in Fig. 5-6 as a function of time in years for $E^2/Q = 10,000$, or directly as a function of $E^2/Q\sqrt{T}$. It would not be too much more difficult to calculate the temperature distribution for any arrangement of wells. Observe that the thermal evolution is much slower than for the first hypothetical model (nonconducting walls). This corresponds to the fact that in the long run, taking account of the conductivity of the rock, we can utilize the heat in an increasing height of the ground. However, a precise study for a particular case should take the thickness of the aquifer into account.

Thus, in a relatively fair manner, we can predict the long-term evolution of a well's temperature and thus the commercial conditions for operation. If a uniform movement, even if it is very slow, exists in the subterranean water, it can serve to assure the permanence of the heat supply. It is consequently very important to document such movement by careful measurements of the static pressure, so as to best choose the locations of injection and extraction wells. It has earlier been seen that viscosity variations, which were neglected, tend to benefit the circulation of warmer waters, which will be favorable for extraction of heat.

But the permeability of the ground is not necessarily uniform; far from it. This nonuniformity can lead to a deformation of the current filaments, but more important, a large proportion of the extracted flow will originate from a small number of filaments. The cooling in those places will be rapid and will have a great influence on the temperature of the extracted water. When it is a question, as we supposed previously, of a sedimentary layer, we can expect that the fluctuations in permeability will be small. But if we imagine a fissure, essentially plane and extended, to which we try to apply the previous reasoning, it must be remembered that the fissure's width is not uniform but leads to the establishment of preferential circulation routes. These routes cool much more quickly than the rest of the fissure and have an important effect on the temperature of the extracted water.

5-9 Geothermal energy prospects at low temperatures

The utilization of thermal energy at low temperatures (i.e., outside the limited zones of abnormal heating) opens up large possibilities. Although this is a favorable circumstance, high values of the heat flow or the gradient are not a necessity. But it appears that deep aquifers which are sufficiently permeable are absolutely necessary. Certain of these aquifers (for example, in the Hungarian basin or the artesian basin of Australia), are supplied with a sufficient amount of water to allow a simple exploitation. In other cases, injections of water will be necessary either to maintain the pressure or to avoid pollution. But the existence of a sufficiently permeable layer seems to be the essential condition of such an operation.

It has been proposed by NASA (National Aeronautics and Space Administration) to attempt heat extraction from a massive impermeable rock, such as a granite, found at a sufficient depth by producing fissures in this rock by the technique of *hydraulic fracturing*, extensively practiced in the petroleum industry. A fracture produced in a first well would be intercepted by a second well, and a forced circulation would be established with the hope that the thermal shock would lead to secondary fissuring, which would enable the rock to liberate an appreciable amount of heat to the water circulating between the two shafts. Such a project involves a number of unresolved problems. Will the fissuring really reach an appreciable proportion of the rock? Will it not establish a group of preferential routes where the major portion of the flow will pass and whose walls will consequently cool rapidly? It would be premature to think that these difficulties can be resolved.

It has also been proposed that subterranean fissures be created by means of nuclear explosions. In fact, even now little is known about how fissures could be created by such a method, and it is almost certain that they would be very irregular with major portions of the circulating water passing along preferential routes which would be rapidly cooled.

5-10 Prospecting

Before being able to exploit a geological site, whether it is geothermal or mineral, it is necessary to establish its existence. First, a favorable zone is identified and places are specified where the existence of an economically feasible geological site seems probable. Then, its existence is proved by sinking drill holes, and the size of the site is established.

This prospecting is actually the most delicate phase of

development. There are too few known examples to allow rules to be formulated, and it would undoubtedly be an error to define a routine procedure to direct such prospecting. In addition, we have seen that geothermal sites are of relatively recent genesis. This means that we cannot count on erosion having exposed parts of the site or its characteristic alterations.

The use of geothermal energy began with already known geothermal sites. At Larderello, wells were sunk to obtain steam in order to increase the natural steam supply of the field and consequently its supply of borax. In Iceland, geyser activity was well known and this knowledge was sufficient to tap these sources at depth.

The case of New Zealand is particularly interesting. Certain thermal zones displayed very spectacular activity, and the Maoris were already utilizing this activity at certain vents to cook their food. Domestic uses were developed in the neighborhood of these zones. When the government of New Zealand attempted an industrial production of geothermal energy, extremely extensive studies were carried out and the results of these studies were published. At first, interest was in surface zones of activity, and measurements of the superficial thermal flux that they represented were performed. But in retrospect, it is apparent that the heat flow which could be obtained underground had no correlation with surface manifestations. The fact that at Wairakei the western extension of the site, which seems the most productive, gave rise to no surface indications and that, of all the fields which were prospected with wells, Broadlands, which seems the most promising field, gave surface indications which were much less spectacular than other fields having less promise, led to a great reduction in the importance attached to surface manifestations. At the same time, in Italy the site of Mount Amiata was discovered, which showed surface thermal sources at very low temperatures only. The California site (the Geysers, where there were no geysers but only a few fumaroles, and the Salton Sea, where there were almost no indications) confirmed the conclusion that there is no strict correlation between the value of the geothermal site and the superficial indices. Should we, inspired by the ideas of petroleum geologists, go so far as to say that superficial indices show only that the geothermal site is being degraded by losses? This would undoubtedly be going too far, but it is certain that these indices (i.e., mud pots, fumaroles, etc.) must be studied in a regional and not strictly local perspective.

Of course, all the methods of geology will be applied to try to specify the local structure. But these methods are often foiled if it is necessary, for example, to determine the structure at depth of a recent sedimentary basin and the positions of permeable levels and impermeable layers capable of forming covers. Geophysics can furnish some complementary information. Without furnishing a foundation for direct prospecting, it should be observed that the electrical resistivity of a soil which is saturated with water diminishes as the temperature increases. But all these methods only provide an outline. Hydrogeologic knowledge of the area is essential, as much to predict if there exist isolated aquifers which could be locally heated as to analyze surface perturbations, particularly the infiltration of cold water which can lead to disturbances of thermal-gradient measurements. The study of this system implies the listing of all pertinent sources and the examination of water flows (which can infiltrate into the ground or be reinforced by subfluvial sources). The flows will be determined as precisely as possible and obviously all the temperatures will be measured. Of course a systematic geochemical study is required. As we have seen, the silica content directly indicates the maximum temperature reached by the water, and variations of other constituents can draw attention to the relationships between waters of different origins and give some information about these waters.

If a geothermal site exists, it necessarily gives rise to a thermal flux by conduction which should be revealed as an elevated thermal gradient. An essential part of prospecting is trying to determine the gradient.

Unfortunately, when permeable soils which can be infiltrated by water exist at the surface, the aforementioned

thermal gradients can be totally perturbed and their measurement will only be valid at a sufficient depth. The pros-
pector should place a high value on the study of the near-
surface materials from which one attempts to determine
where, and at what depths, it will be possible to measure a
representative thermal gradient in a drill hole. In uniform
soil, impermeable and saturated with water, wells should
be sunk for thermal-gradient measurements to a depth of
30 to 50 m in a grid pattern, established a priori, which will
allow the places with the highest values of the gradient to
be found. However, with permeable, aerated, upper layers,
and where cold surface water seeps in, the results are not
necessarily significant; and it is necessary that deeper borings be undertaken to measure the gradient below the perturbed zone.

In such a campaign to measure the temperatures,
should we sink several tens of costly drill holes to depths of
around 50 m, or should we plan temperature measurements
in holes of about 2 m, which can be dug by a laborer in soft
soil in less than an hour and lined with plastic tubing? It is
certain that for many fields the gradient is such that a significant variation can be seen at the shallower depth. But
we should only compare measurements corresponding to the
same surface conditions and observe the long-term (annual)
temperature variations in several test holes. Obviously
these measurements must not be done until a certain waiting period has elapsed. As a precaution and in the absence of
unduly marked superficial perturbations, such a campaign,
called a preliminary reconnaissance, can be justified. This
does not mean that we can dispense with the prospecting
phase where temperature gradients are measured at a
depth of around 50 m.

When these gradient measurements indicate a hot subterranean region, it is still necessary to consider if the geologic formations allow us to hope to find a permeable layer,
suitably supplied with water and possibly protected by a
relatively impermeable cover. But these data and what
they imply about the existence of a geothermal site can be
further specified only after one or several wells have been

sunk with the same precautions as for commercial wells (possibly with smaller diameters) and lined with a casing before making any measurements. A reliable thermal profile can only be obtained after a sufficient waiting time, but more hasty measurements can be useful to direct the course of the preliminary reconnaissance.

To what depth should we sink the wells? Field production experience shows that a depth of 500 m is considered normal. At lesser depths, we have only weak geothermal sites. There should be no hesitation to sink deeper drill holes if there is some indication that the temperature is continuing to increase. At the Geysers, deeper wells discovered a more extensive site than one that was exploited earlier at 300 to 400 m. Thus, it seems reasonable to consider the possibility of sinking reconnaissance wells to 1,000 or even 1,500 m. Obviously, there is no set rule.

Once the thermal profile is measured and after a sufficient waiting period, if high temperatures have been found, it is necessary to try to make the well erupt. If this does not occur spontaneously, then it is necessary to cause an eruption by reducing the head of water in the well with a pump, or, even better, with an air lift. As it is not very well known to what extent the cold mud used for drilling has penetrated into the ground, we should wait until the eruption has reached a steady flow before taking production measurements (their purpose was indicated earlier). It would be imprudent to undertake a commercial project before having the results from several wells, of which some will be of the same diameter as the commercial ones, and having let these drill holes flow for a long enough time to assure the permanence of the flow.

Finally, the decision to exploit a geothermal site is not only a technical problem. It also involves economic aspects which we will briefly examine in the next chapter.

general remarks on the economic management of geothermal energy

6-1 General view

If, in comparison with the operation of a conventional thermal power station, the savings realized by not buying fuel are considered and if we add the savings resulting from the absence of a boiler against the drilling costs and equipping of wells for steam extraction, it appears that geothermal power should be more economical.

But on the other hand, it is necessary to consider the fact that geothermal generator groups are small units with turbines operating at low pressures often dispersed over a rugged terrain, these groups are often distant from rivers which would permit efficient cooling of the condensers, and the construction of cooling towers may be required. Also, prospecting is not guaranteed to succeed, and if an exploitable site is not discovered, the money spent will be lost. These are the unfavorable circumstances which interfere with profitability.

A return price per kilowatt-hour could be calculated[1] for a geothermal plant and for conventional plants taking account of capital investments, amortization at a rate which must be established, and operating costs. But if these items are examined a little closer, it is realized that such a calculation does not have a great deal of significance because of the uncertainty of the various factors and also because the circumstances depend on the economic situation in which the new power station is placed.

[1] For example, see G. Facca and A. Ten Dam, "Geothermal Power Economics," Geothermal Exploration Company, Los Angeles, 1964.

The ideal case would be where the power plant only has to supply a limited number of consumers, such as was the case at Kiabukwa. In such a monopoly situation, it suffices to fix the sales price for the energy in order to amortize the expenditures, cover the operating costs, and realize a certain profit.

6-2 The marginal economic situation

However, the situation is rarely so simple. When a new producer of energy appears, in general there already exists a distribution network which satisfies the requirements of the consumers and it is through this network that the energy supply must pass from the new power station. What should be the price of the energy sold to this distribution network?

In the case of a classical liberal economy which considers separately the costs affected by the decision to buy this new energy, envisaged as marginal, the answer may well not be at all the same as if the problem of energy production in the long run is considered. The distributor will fix the buying price for the geothermal energy which is offered to him by calculating it not with respect to the total cost of the energy produced by the existing power stations, but with reference to the marginal cost (i.e., the savings realized by not producing that power). In the cost of a produced kilowatt-hour, expenditures of different natures occur; certain are proportional to the energy produced (fuel costs) and others are constant in time (security, maintenance). Finally, on the one hand the capital invested in the construction of the power plant must be paid back, while, conversely, capital must be put aside for the day when the power station, becoming antiquated, can no longer operate. But this amortization does not occur only for the power stations which produce the energy for which we seek the price. The demand—daily, weekly, annually—varies strongly from the peak hours when demand is maximum and to base hours when consumption is much lower (the base hours refer to the power which is constantly utilized). Due to existence of peak hours of consumption additional plants must exist to supplement, if only for a few hours per day, the normal energy production. Nonetheless these extra plants must be amortized. Finally, the distributor guarantees the power supply required in spite of possible problems which might occur either due to the consumers or to production problems (breakdowns, repairs); this obliges the distributor to have a certain production capacity in reserve, which is never used except in emergency (or which is operated in rotation) and whose amortization and fixed costs must also be charged to the cost of the energy sold.

For the producer, the difference in value of the energy at peak or base hours may be large enough to justify the construction of regulating works which store energy during base hours (by pumping water into an elevated basin) for use later during peak hours, in spite of the cost of such installations and the inevitable energy losses.

When a geothermal power station offers its produced energy, it appears obvious that no savings are made by not producing energy during base hours. Or more exactly, the saving realized in leaving the nonextracted heat in the ground will only materialize at the end of the operating period by prolonging its duration. This is a nebulous term which is too uncertain for economists to be disposed to take it into account. Thus, the production of the geothermal power station will be utilized to assure the base energy supply, energy to which the producer gives the lowest possible value (the same problem occurs for nuclear power plants).

The distributor will perhaps try to offer only the marginal cost of the base energy on the pretext that his equipment allows him to guarantee the power required. But the geothermal station should figure in the power reserve which is necessary if the distributor is to guarantee service, and the price of the energy which the station produces should reflect this. A large network, facing a constantly increasing demand, must always have supplementary power in reserve so that it has the margin which will permit it to give its guarantee. Thus the existence of a geothermal power station permits the distributor to dispense with the construction of an equivalent power station, and the buying price of the energy should reflect this situation and cover the corresponding amortization over and above the marginal cost savings. But this is less evident for a small network which can estimate that it already has the power necessary to guarantee its supplies. The distributor may not be disposed to consider the time when it will be necessary to add to its present equipment; such a project is delayed, and the distributor can thus avoid, for a time, gathering the financial reserves which will be necessary for these additions.

We are often placed in the situation—very frequent in actual practice—where the producer of geothermal energy is distinct from the distributor through whom the energy must pass. The latter has a quasimonopoly which he may be tempted to abuse. This situation would not be changed if, instead of selling the electrical energy produced by the plant, the geothermal plant operator sold the steam instead. This was the case at the inception of the Geysers' fields. However, since the steam flow is difficult to measure, it was finally agreed to calculate the price according to the electrical energy consumed.

It can also happen that the distributor, anxious to diversify his production sources, will undertake a geothermal energy operation of his own, which would permit a more equitable valuation of the distribution of the return price.

Implicitly, we have assumed a situation such as in Western Europe or the United States, where all the large industrial states have an extensive interconnection and where the energy consumed is very large with respect to the increase of energy made possible by the use of geothermal power. Because of this fact, there is no need to consider the effect of geothermal energy production on the amount of energy consumed.

It is not the same situation in an isolated region, and it is in such conditions that we may hope the production of geothermal energy will have a marked influence on the development of a country. The volume of energy consumption will then strongly depend on the price at which the energy is sold. There are certain needs (domestic lighting, air-conditioning, refrigerators) for which consumers will accept any price that is demanded of them. A second category of consumers will compare this price to the cost of another solution such as producing their own energy with a diesel. Again, it would be necessary to calculate the value of the guarantee resulting from connection to the network, even without a normal call for energy.

Schematically, a third consumer category can be imagined—those industries which have installed their plants to take advantage of the price at which energy is offered to them. This only applies to industries which are large consumers of electricity, such as those that prepare aluminum from the electrolysis of alumina. The localities of certain plants—in Norway or British Columbia—are only justified by the availability of hydroelectric energy at very low prices and in large quantities. It can be said that the long-distance transport of alumina and aluminum is the most economical method of transporting electrical energy beyond the relatively limited distances permitted by interconnecting power networks.

When considering the development of geothermal energy networks in underdeveloped regions with a low consumption rate, it is naturally thought that the presence of the power will attract new industries. But there can be a considerable difference between the power which can be produced on the spot and the energy which must not only be produced but must also be guaranteed for a sufficient time to attract new industries, which require, besides low-cost energy, other resources such as water, a qualified labor force, transportation facilities, etc.

6-3 General remarks

Can the attitude of a traditional liberal economy which considers a decision such as the construction of a geothermal power plant in terms of immediate financial objectives be justified when formulating longer range goals?

The economic basis of our civilization is supplied by nature in the form of energy resources which have been developed by us and distributed in the form of fuel for mobile engines or electrical energy. It is scarcely necessary to recall the successive stages in this energy evolution: domestication of animals; development of windmills, sailing vessels, and water wheels; development of steam engines fueled with coal first for local power production and later integrated into electrical power systems; transformation of water wheels into hydroelectric plants and, at the same time, regulation of energy production for peak-energy demands through the use of dam reservoirs; partial substitution of hydrocarbons for coal first in mobile vehicles (where they have gained a monopoly) and then in fixed power-plant locations; and use of easily distributed natural gas. Most recently, nuclear energy has appeared, which at present—and perhaps for a long time in the future—is confined to large fixed installations which, for technical reasons, have to be used for assuring base-power requirements. Even in the realm of nuclear energy it is necessary to distinguish between "classical" fission-power stations, which have been technologically developed to the point where almost identical installations can be constructed, and the experimental prototypes, such as the breeder reactors, which will use the natural resources of uranium more economically in the long run (this does not mean to say financially more economical in the short run). This says nothing of the hope, perhaps distant and illusory, of a domestication of fusion energy.

Even such a brief survey makes it clear that there is a constant increase in global energy consumption and that the means of producing it must also be in constant evolution. This evolution would be even more striking if we were to enter into the details of production techniques and analyze, for example, the evolution of steam engines from the first piston engines up to high-pressure turbines, constituting more and more powerful units, or those evolutions which led from the internal combustion engine to the jet engine and the gas turbine.

If one of the factors in this technical evolution is the constant problem of increasing the efficiency of production, another factor results from the availability of natural resources to exploit. Coal mines fed the essentials of the industrial revolution of the nineteenth century, but in a country like France, their future seems very limited for two reasons which perhaps have not been sufficiently distinguished: their price is higher than that of imported hydrocarbons and the reserves are depleted. The two factors are interdependent because the reserves can only be estimated if the maximum price that one is prepared to pay for their extraction is fixed.

The problem of reserves and the depletion of resources is particularly acute for hydrocarbons because of the ease and thus rapidity of exploitation once they are discovered. With a doubling every 20 yr in consumption and with reserves which scarcely can last more than 20 yr, prospecting ought to be constantly increasing, although the virgin regions where exploration can be carried out are rapidly decreasing. Prospecting has occurred recently in shallow marine waters. Undoubtedly this exploration will be extended to deeper water, but the time when all these possibilities of extending the areas of exploration are used up is inexorably approaching.

It can be added that oscillation between two extreme political positions can be seen. One of these positions considers that a market economy for a worldwide production of energy will furnish the most economical solution, and the other viewpoint stresses the advantage of dependence on national resources whose availability is not subject to international tensions.

Finally, another consideration appears, the importance of which some people have tended to exaggerate. Certain mineral operations degrade the countryside in which they are carried out. An example is strip mining. In addition, the utilization of sulfur-bearing fuels introduces a gas into the atmosphere which normal respiration can tolerate only in infinitesimal doses, and which now risks reaching excessive concentrations around some industrial centers. Nuclear reactors leave waste products consisting of highly radioactive fission products the existence of which constitutes a permanent menace. It appears necessary to take these pollution factors into account, and that could seriously limit the development of certain energy sources.

In this picture of constant evolution and sometimes somber perspectives, what place can be envisaged for geothermal energy? Its present influence is extremely small, practically limited to a few particular cases. The operation at Larderello (Italy) happened to facilitate the electrification of the railroads, but the percentage of energy which it gives to the Italian energy pool is only 3 or 4 percent. It is the same situation to a much greater degree for the production of the Geysers, which is fed into the Western United States energy grid. In New Zealand, the development of geothermal energy could technically go much farther, but the increase in consumption is not large enough to justify it. New Zealand has abundant hydroelectric resources, chiefly on the South Island. The laying of an underwater cable (600,000-V direct current) in Cook Strait has retarded the development of the next geothermal field (Broadlands) for several years. However, I believe it is necessary to emphasize that this delay has no relation to the economic success of the operations at Wairakei but arose essentially from the structure of the energy market in an isolated region.

Outside of the case of New Zealand, where the geothermal potential is a part of the energy reserves which will be developed as the need arises and will guarantee consumer satisfaction in the future, and perhaps Iceland, where the availability of heat at low temperatures at a cheap price for domestic heating and greenhouses has a significant importance for the country's development, what is the future of geothermal energy?

It is necessary to emphasize one of the reasons why it is difficult to answer this question; we are only at the beginning of geothermal energy production. At first, this development occurred at a few exceptional sites which we are sure exist only in a very few countries, and it is only more

recently that geothermal fields having real economic value
and which displayed few superficial indications were discovered. It is these latter discoveries (which now exceed 10, although most of them are not yet in full commercial operation) that we should consider as a basis for estimating the
future of geothermal energy, and not the first generation of
plants constructed on exceptional sites which were evident,
such as Larderello and Wairakei.

In addition, the understanding of the phenomena
which act in geothermal beds, to which we have devoted
Chaps. 4 and 5, is still very imperfect. There remains much to
be learned about the way in which a pressure drop can cause
vaporization underground and in the way in which the
steam can escape, either by means of the permeability of the
rock or through a network of fissures. Also the technology
utilized for geothermal exploitation seems extremely rudimentary; numerous improvements could be considered,
from the enlargement or even the creation, of fissures
through which the ground can be drained to the selective
exploitation of steam by drill holes. Consideration of the
evolution of oil-field technology in one century leads us to
imagine what a comparable development could bring to
geothermal operations.

If this technological progress is achieved, if an inven-
tory of geothermal resources is actively encouraged and
methods of prospecting are developed and perfected, I think
that a prudent appraisal of the future of geothermal energy
would permit us to say that at least in certain countries (the
imprecision of this phrase reflects the prudence which must
be imposed) the utilization of geothermal energy will be able
to attenuate the energy crisis that the depletion of hydro-
carbon reserves could make acute in 20 or 30 yr.

Attenuate, but not mitigate, because different forms of
energy cannot be casually substituted for one another.
Geothermal energy will be exploited by fixed stations at
sites determined by the availability of hot water or steam
for the basic purpose of assuring base-power consumption
(although a daily regularization by accumulation of steam
in subterranean reservoirs is not inconceivable). The role of

these power stations will thus be analogous to atomic power stations, which appear to represent the ultimate response to the energy crisis in the long run. But, it is still probable that the crisis will be especially intense for mobile vehicles, which are almost exclusively dependent on hydrocarbons.

Yet, it seems to me that it would be imprudent to count on nuclear power stations entirely and to renounce the possibilities that geothermal energy offers in certain regions and in particular for isolated regions which cannot claim a priority for the construction of nuclear power plants. But in order to obtain the geothermal energy which ought to be expected, at the moment of need and in economic conditions which could be quite different from those of today, it is necessary between now and then that research be continued, tending on one hand to identify geothermal sites (even if their exploitation must be postponed) and on the other hand to improve the operational technology which will be required.

It is necessary, it seems to me, to specify the aim in this development of technology with care. We can decide to undertake an experimental operation to test the capacity of a geothermal site, and the permanency of its production and to try to bring technological improvements to the methods of exploitation. Energy production in what could be called a pilot plant will result from these efforts. Of course, attempts will be made to sell this energy under the best possible conditions, and that will help defray the total costs of the operation of the pilot plant. Indeed it may entirely support these expenses, even if the amortization of installations and exploration cannot be assured.

If we leave aside the future of the power stations of New Zealand and the Geysers, for which the profitability under present conditions seems already assured, it would seem to me that this perspective of the development of pilot plants, which are technically indispensable if the long-term ability to construct economically profitable power stations is desired, must be kept in mind while making all decisions to undertake the exploitation of a geothermal field.

The first power station at the site of the Geysers, with 14,000 and then 27,000 kW, was in fact nothing more than a pilot plant and would undoubtedly never have allowed the amortization of operations performed prior to its installation. But this represented a necessary experimentation stage before a larger production capacity could be considered.

By definition, the profitability of a pilot plant is not assured in itself. To estimate the economic viability of future developments which it makes possible, even in the long run, it would be necessary to take into account the evolution of the site on which the plant is installed, as well as the value of technological improvements that can be hoped for and that may find applications elsewhere. It is with this perspective, and not that of marginal profitability on the actual energy market, such as was analyzed in this chapter, that all decisions concerning the development of geothermal energy should be made.

6-4 The legal regulation of geothermal energy

Perhaps we should mention here a necessary condition for the development of geothermal energy, although it is certainly not enough in itself. A convenient legal system, inspired by mining law, must be established to assure the explorer the benefits of his discoveries, without making him liable to unreasonable demands from the land owner, and to protect him from competition from other groups who would like to take advantage of his discoveries to operate on the same site. A simple extension of the mining law, differing from one country to another, could in general protect prospectors in these situations. In France, this extension was made for the Antilles and the French territory of the Afars and the Issas, precedents on which an extension of the same concept to the metropolitan territory shall rely—an extension which must occur in the very near future. In fact, such an extension will simplify the legal system while assuring the protection of miners' exploration rights in low-temperature fields (low-temperature fields means less than 100°C). The New Zealand system could provide a model for Anglo-Saxon mining law. In the United States, a law went into effect on December 29, 1970, which defined the conditions under which exploration and exploitation rights could be given for public lands (i.e., the quasi totality of the land surface in the Western United States, where the federal government reserved the mining rights for itself when other land rights were (if they were) given to others). Although the interest in geothermal energy operations that arose following the success of the Geysers fields could not be manifested until this law, geothermal energy enterprises, some with shaky foundations, arose and suddenly flourished and much interest was manifested on the part of private industry and by diverse public organizations in geothermal energy operations.

The legal system instituted for the Western United States has three variations, subject to more and more rigorous conditions of competition, depending on whether the regions considered correspond to zones where the presence of geothermal sites is considered as "possible" or "probable." However, the U.S. Geological Survey had to make this classification in a very brief time for a considerable area of land, over 2 million km². No prospecting was possible in the limited time available, and the classifications which were given rest on some very tenuous presumptions. Possible serious misunderstandings can occur if the word "probable," which only designates a certain procedure of distributing licenses, is confused with the sense of that word when referring to mineral reserves.

6-5 The role of the public sector

The case of a new source of energy, which appears likely to become economically competitive at the end of a more or less prolonged period of development, is not new, and it is interesting to analyze some examples of how such developments were financed. From Denis Papin to Fulton, the development of the steam engine and its applications was in the main left to isolated inventors who tried to convince capitalists with a sense of the future to help them. With multiple episodes, sometimes with heartbreaking checks and delays, this system functioned better than worse, although it is evident today that it is outmoded. We could also point out the birth of the internal combustion engine, whose inventor was not financially rewarded, as his actual success would have justified. (In fact, he was not rewarded at all.) A majority of other developments, for example, the steam engine and the appearance of the turbine, were done progressively and were financed as work continued by constructors, thanks to immediately realizable increases in efficiency.

The jet engine of aviation required enormous expenditures on research, experimentation, and development; it was entirely supported by the military, i.e., by the public sector, and it matters little to us whether the justifications for these expenses were economic or strategic.

Since its beginning, nuclear energy also benefited from military-oriented financing. It was clear in 1945 both that there was a possibility of nuclear energy production and that it would only become economically competitive at the end of a long development process. This development was supported by the public sector, but not without the intervention of strictly military considerations at different stages. Recall the progress of this development from low-power pilot plants to powerful reactors which were not economically competitive until the present time. Today the extrapolation of economic conditions over a period of 20 yr for different possible energy sources amply justifies the construction of nuclear power plants.

One also knows of the efforts devoted to the breeder reactor, which no military considerations now justify. It is not even certain in the immediate future that these devices will be economically competitive with the slow neutron reactors already in use. But its best justification is in the long-range savings of uranium, the known supplies of which today are minimal with respect to future needs. The interest in the breeder reactor in one of the rare examples of a case in which an actual decision is affected by considerations of foreseeable conditions 30 to 50 yr in the future.

This brief review was designed to show that the actual development of a new source of energy cannot be left to private financing alone. Funds to develop new ideas, or for improving installations which are already economically justified, are unlikely to be adequate. It is apparent that such development is a political problem.

In what countries will this development take place? Should the progress already realized in more advanced countries (which in our case includes New Zealand, Italy, and the U.S.A.) be utilized in other countries by acquiring licenses or employing foreigners who are experienced in geothermal energy equipment and production? For countries of moderate industrial importance such as France, such solutions were not considered for the jet engine or for nuclear energy. Obviously, military considerations influenced these decisions, but in retrospect it appears that independent development was a necessary condition for the ability to participate afterward in information exchanges and the international cooperation which results from such exchanges.

Even though absolutely no military considerations are relevant to the case of geothermal power, the influential countries whose territories offer some possibilities in this realm are undoubtedly interested in financially encouraging its technological development, from adaptations and improvements realized in other fields, to assure the best utilization of their natural resources.

6-6 The immediate effects of the increased price of energy

The large increase in fuel costs, which must be interpreted as the end of a transitory era of cheap imported fuel, led to a reexamination of the overall energy market. In Western industrialized nations this market is divided into three approximately equal divisions corresponding to transportation, industry, and heating.

For this last item, the fuel (oil, gas, coal, or even electric heating energy) produces calories that are utilized at a temperature only slightly exceeding the ambient. If geothermal heat were directly employed, it could easily substitute for these other resources calorie for calorie. Thus, for geothermal energy at low temperatures (which may be much more available than is currently contemplated) there is a very important future.

However, there still exists a technological problem: our habits in the matter of heating rest on the ease of heat transfer with large temperature differences. Without considering the use of fireplaces, central heating consists of a boiler, the surface area of which can be quite small thanks to the high temperature of a flame, and radiators at a temperature of perhaps 90°C to heat a room to 22°C.

The utilization of geothermal energy requires more efficient heat exchangers with only slight temperature differences. There is first the exchange between the extracted groundwater (which must generally be reinjected, if not to reduce pollution, to maintain pressure in the geothermal bed) and the water of the heating circuit, and afterward between the heating circuit and the area to be heated. Heating elements in the floor, which can be used at 30°C, are one possible solution, but there are certainly other ways to improve the heat exchange between radiators and the surrounding air.

From now on, in areas of known deep subterranean aquifers, new housing could and should be constructed to utilize geothermal energy as the heating source. The renovation of older housing poses a more delicate problem. It is probable that boilers will be retained for exceptionally cold periods. We could also imagine a combination of heat pumps with geothermal heat which may be justified by the fact that it is sometimes costly and difficult to find an adequate cold source for conventional heat pumps. Subterranean geothermal waters, even at only a slightly elevated temperature, constitute such a cold source and allow heat pumps to supply conventional radiators at the necessary temperature.

Thus, the problem is to reconsider heating installations and to disseminate new innovations. Aquifers should also be sought, which means recognizing permeable areas in sedimentary basins. In France, we could cite the sands of Lussagnet in the south of Aquitaine, the sandstone of the Trias in the north of Rhine graben, and the base of the Oligocene strata in the north of the area around Limagne. Obviously, this list is not comprehensive. Undoubtedly, other countries offer similar opportunities.

By such efforts, even though we cannot solve the energy crisis, we can save several million tons of fuel per year in the next few years. This prospect amply justifies the necessary efforts.

thermal aspects of some geological phenomena

It will not be possible here, even in a quick review, to explain volcanic activity, tectonic deformations, which imply a dissipation of energy which is essentially transformed into heat, or metamorphism, the fundamental cause of which appears to be an elevation in temperature. But in the study of such problems, properly thermal questions are incidentally encountered, and it seems useful to indicate some solutions.

7-1 Volcanic eruption mechanisms

Volcanism implies, in the first place, openings in the rocks of the crust through which products at high temperature rise and flow out at the surface or sometimes infiltrate along fissures and joints in the ground. These products are essentially composed of a molten silicate magma which, upon solidification, yields a lava but can also contain an appreciable number of volatile substances in solution (water, CO_2, SO_2 or H_2S, etc.). These gases are sometimes released in abundance, but we cannot say whether they existed in a free state or only dissolved in a magma underground. One feature of volcanism is its extreme irregularity in time, whether on the scale of an hour, day, year, or century.

The first question to ask concerning volcanism concerns the piercing of the crust by molten lava, generally very viscous, which rises in a fissure or moves horizontally along a bedding plane to form a sill or laccolith. A hydrostatic interpretation, taking into account the differences in density between molten lava which is lighter than the rocks of the lower crust and this same solid lava and the lighter sedimentary rocks, often permits us to clearly visualize the distribution of such injections. But this does not indicate the piercing mechanism, which requires a considerable excess of pressure to break the rocks in the vicinity of the injection. Now these molten lavas are very viscous, which means that when they advance there is a substantial pressure difference between the leading and rear portions, and it is difficult to understand how the increase of pressure, which is capable of breaking the surrounding rocks, is produced. The solution of this paradox is undoubtedly thermal. At a given stage of the injection, heating of the surrounding rocks vaporizes the water which these rocks contain. The steam reaches a pressure in excess of the pressure which is acting on the rocks and consequently shatters them. The steam then escapes through the fissures thus formed and condenses on contact with cooler rocks thereby reducing the pressure in the pocket which the steam has opened, and the viscous lava can progressively occupy this space. The same cycle then repeats itself. When a high vapor pressure persists at the leading edge, the lava cannot advance, but it is too viscous to be pushed back.

7-2 Cooling of a volcanic injection

Geologists often pose the questions of the time necessary for the cooling of such an injection (or flow) and the temperature attained by the surrounding rocks at some distance. These questions would be easy to answer if the thermal characteristics of the lava and the surrounding rock were exactly known. This is not the case because the thermal conductivity can change with the temperature, generally slightly decreasing while the heat capacity increases, often by 50 percent at 400°C. Since we do not know the details of these variations, we must be content to use average values which are more or less happily chosen. The results of these calculations, which we will indicate, will only have the precision of the given values, but in any case we will obtain valid orders of magnitude and a description of the general behavior of the phenomena.

The thermal perturbations are imposed on a prior thermal system composed of a steady thermal gradient in equilibrium with the geothermal flux. It is this prior temperature distribution which must be taken as zero. Thus, the temperature of the injected lava will be calculated as an addition to the initial temperature of the rocks, and the calculation of the heating of surrounding rocks will also be taken from this temperature. It is clear that the thermal perturbation must satisfy the heat equation.

The heat carried by a molten lava, which will diffuse in the surrounding rocks as the cooling occurs, is partly heat corresponding to the specific heat of the lava and partly the latent heat of crystallization which, for a pure material, is released at a constant temperature during solidification. But we do not know if the injected lava was already carrying crystals, and we are also not sure that the last phase of crystallization occurs at a constant temperature. We will indicate the principle of the rigorous calculation including the latent heat. However, this calculation sometimes raises certain difficulties, and it is often more convenient to neglect the latent heat. To compensate we may slightly increase the temperature attributed to the injection, and consider its cooling with a constant specific heat.

It remains to take account of the way in which the injection occurs. A dike can be emplaced in several distinct ways. It can be made all at once by the injection of basalt in a fissure which opens rapidly. Or it can be made progressively, with new basalt being injected between two older layers of basalt which are plastered against the fissure walls. The old basalt is cooled at its contacts with the country rock, while it is heated by the new basalt toward the axis of the dike. But it is rare that an examination of a dike allows us to choose between these two interpretations. Even if we accept the second, the total injection time remains unknown. It may have been brief with respect to the cooling period.

If we wish to take latent heat into account, we will consider the wall of a dike brought, at the moment $t = 0$, from an initial temperature θ_0 to the molten lava temperature θ_1 and remaining at this temperature until all the latent heat is dissipated. If the thickness of the lava dike is $2e$ cm, the latent heat L cal/g, and ϖ the specific gravity, the heat which must be dissipated before the lava is entirely consolidated is $e\varpi L$ cal/cm². We have seen that the temperature at distance x from the wall will be given by

$$\theta = \theta_0 + (\theta_1 - \theta_0)\, \text{erfc}\left(\frac{x}{2\sqrt{\alpha t}}\right)$$

The heat flux at the wall is given by $K(\theta_1 - \theta_0)/\sqrt{\pi \alpha t}$, and its value integrated in time from the origin is $2K(\theta_1 - \theta_0)\sqrt{t/\pi\alpha}$. These equations lead to the total time for the solidification of a lava dike:

$$t = \frac{\pi \alpha e^2 \varpi^2 L^2}{K^2(\theta_1 - \theta_0)^2} \tag{7-1}$$

Observe that this time is proportional to the square of the thickness.

We have implicitly assumed that the temperature in the dike was uniform and that the heat exchanges took place easily enough so that the latent heat was dissipated from the dike surface. This would be reasonable if the lava remained liquid until solidification. But it cannot be the same case if the lava is almost solid, and thus the latent heat must be dissipated by conduction. In this case, there would be a progressive solidification from the walls with a thinning of the central fluid core, and the displacement of the solidification boundary x_L would be tied to the gradient in its neighboring solidified side by

$$\varpi L \frac{dx_L}{dt} = K \frac{\partial \theta}{\partial x} \tag{7-2}$$

Hence, it is evident that the progression of the solidification does not depend at all on the thickness of the dike and the central portion that remains fluid but only on the distance from the wall where the cooling occurs.

We can easily find the temperature law applicable to the surrounding rock and the solidified part of the dike, the temperature θ_1 remaining constant in the remaining liquid portion.

This temperature law is identical to the law which would result from the cooling of a dike, supposed initially at a higher fictive temperature θ_2, but without taking the latent heat into account. In fact, in this case, the temperature on the x axis is given by

$$\theta = \theta_0 + \tfrac{1}{2}\,(\theta_2 - \theta_0)\left[1 + \mathrm{erf}\left(\frac{x}{\sqrt{4\alpha t}}\right)\right] \tag{7-3}$$

Let λ be the value of $x_L/\sqrt{4\alpha t}$ for which $\theta = \theta_1$. We find for the corresponding value of x_L:

$$\frac{dx_L}{dt} = -\lambda\sqrt{\alpha/t} \qquad \text{and} \qquad \frac{\partial\theta}{\partial x} = \frac{(\theta_2 + \theta_0)e^{-\lambda^2}}{\sqrt{4\pi\alpha t}}$$

In order for these values to satisfy the boundary conditions, it suffices that

$$\frac{\varpi L}{C} = \frac{(\theta_1 - \theta_0)e^{-\lambda^2}}{\sqrt{\pi}\lambda[1 + \mathrm{erf}\,(\lambda)]} \tag{7-4}$$

From this equation we can obtain the value of λ, with the help of Fig. 7-1, and from that we obtain

$$\theta_2 - \theta_0 = \frac{\varpi L}{C}\sqrt{\pi}\lambda e^{-\lambda^2}[1 + \mathrm{erf}\,(\lambda)] \tag{7-5}$$

This expression can be shown to always be less than $\varpi L/C$.

These formulas describe the temperature evolution until the end of solidification of the fluid core, but not afterward since we must then take account of the cooling through both faces of the dike. The consolidation time of a lava dike of thickness $2e$ cm is given in seconds by

$$t = \frac{e^2}{4\alpha\lambda^2} \tag{7-6}$$

For example, Fig. 7-2, with reasonable values $L = 75$, $\varpi L = 200$, $C = 0.5$, $\theta_1 - \theta_0 = 800°C$, we find that $\lambda = 0.551$, $\theta_2 - \theta_0 = 1027°C$, and $t = 82.7e^2$. For $2e = 2\,\mathrm{m}$, t is about 10

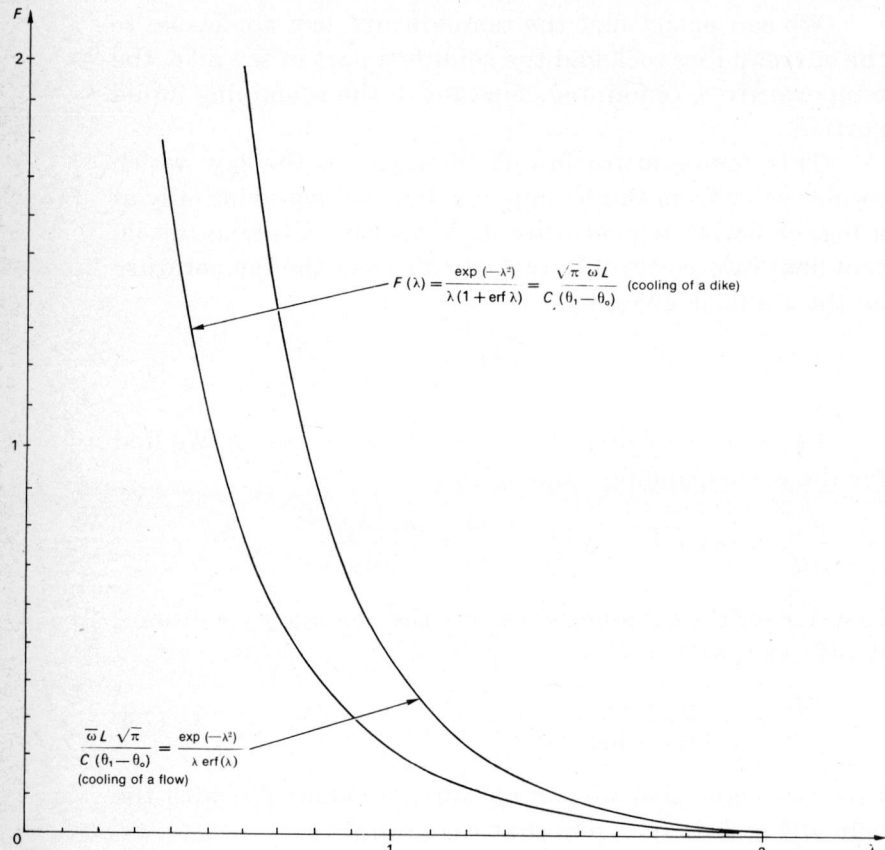

$$F(\lambda) = \frac{\exp(-\lambda^2)}{\lambda(1 + \mathrm{erf}\,\lambda)} = \frac{\sqrt{\pi}\ \overline{\omega}L}{C\,(\theta_1 - \theta_0)} \quad \text{(cooling of a dike)}$$

$$\frac{\overline{\omega}L\ \sqrt{\pi}}{C\,(\theta_1 - \theta_0)} = \frac{\exp(-\lambda^2)}{\lambda\,\mathrm{erf}(\lambda)}$$
(cooling of a flow)

FIGURE 7-1   Graphs for the calculation of λ for studying the cooling of a lava dike or flow, including the latent heat of solidification.

days and until then, the temperature at the boundary of the dike stays at $\theta_0 + 513°C$.

Let us now imagine the case of a dike at an initial temperature θ_1 without any latent heat to be dissipated and with a thickness $2e$. We immediately find, x being the abscissa taken from the middle of the dike, that

$$\theta = \theta_0 + \tfrac{1}{2}(\theta_1 - \theta_0)\left[\mathrm{erf}\left(\frac{x + e}{\sqrt{4\alpha t}}\right) - \mathrm{erf}\left(\frac{x - e}{\sqrt{4\alpha t}}\right)\right] \qquad (7\text{-}7)$$

This formula is general if we agree to the convention that $\mathrm{erf}(-\lambda) = -\mathrm{erf}(\lambda)$.

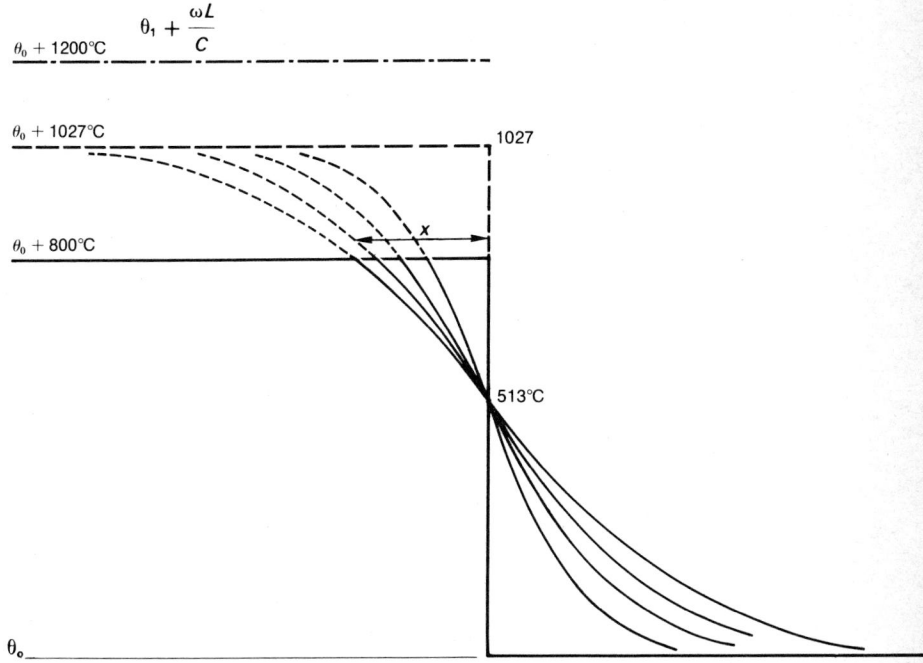

FIGURE 7-2   Successive thermal profiles for the cooling of a face of a lava
dike, including the latent heat of crystallization for $\lambda = 0.551$
$(\theta_1 = \theta_0 + 800°C, L = 75$ cal/g, $\omega L = 200, C = 0.5)$ for times proportional to 1, 2, 3, and 4.

We can also ask the question of when the maximum
temperature is reached at a point on the abscissa $x > e$. We
easily find that for this case it is necessary that

$$\frac{(x + e)^2}{4\alpha t} e^{-(x + e)^2/4\alpha t} = \frac{(x - e)^2}{4\alpha t} e^{-(x - e)^2/4\alpha t}$$

Letting $m = (x^2 - e^2)/4\alpha t$, $\beta = (x + e)/(x + e)$, this equation
becomes $m = 2\beta \ln \beta/(\beta^2 - 1)$ which then gives $t = ex/\alpha \ln$
$[(x + e)/(x - e)]$ and

$$\theta_m = \theta_0 + \tfrac{1}{2}(\theta_1 + \theta_0)\left(\text{erf}\left\{(x + e)^2 \frac{\ln[(x+e)/(x-e)]}{2ex}\right\}\right.$$
$$\left. -\text{erf}\left\{(x - e)^2 \frac{\ln[(x+e)/(x-e)]}{2ex}\right\}\right) \quad (7\text{-}8)$$

Observe that for a given value of λ, this time increases as the square of the thickness. Equation (7-8) gives us the maximum temperature attained as a function of distance.

If the consolidation of the dike brings the latent heat of crystallization into effect, as previously, we can try to take this into account by adding a fictive temperature to the lava's initial temperature. But in the study of the temperature behavior of the surrounding rocks what is important is the total liberated heat, and that leads us to take an additional fictive temperature such that $(\theta_2 - \theta_1)C = L\bar{\omega}$. This is always greater than the earlier fictive temperature which we found, but is of an order of magnitude which is not too different.

The difference between the results which follow from the two types of calculations at the exterior of the dike is undoubtedly less than the effects of surface irregularities of the structure which we did not consider. As a case in point, for the previous example, $(L\bar{\omega}/C = 400, \theta_1 - \theta_0 = 800°C)$, the temperature at the end of consolidation gives a wall temperature of 530°C instead of 513°C, and at the center of the dike, a temperature of 665°C instead of 800°C. However, it is at the center of the dike that the discrepancy between the two calculations is maximal. These results are independent of the thickness. But at the center of the dike, we can actually expect that the molten matter remains in irregular pockets, and not in a band the thickness of which remains constant.

7-3 Cooling of a neck

The conditions for the cooling of a neck, an intrusive cylindrical formation, results immediately from Eq. (1-2). The temperature distribution $\theta = (1/4\pi\alpha t)e^{-(x^2 + y^2)/4\alpha t}$ satisfies the heat equation in two dimensions and reduces, for $t = 0$, to a zero temperature everywhere except at the origin with

$$\iint \theta \, dx \, dy = 1$$

We will take a neck temperature θ_1 at the initial in-

stant, which can possibly take the latent heat of crystallization into account as previously.

At a point X, Y the temperature at a time t will be given
by

$$\theta = (1/4\pi\alpha t)(\theta_1 - \theta_0) \iint\limits_{s}$$

$$\exp \frac{-[(X - x)^2 + (Y - y)^2]}{4\alpha t} \, dx \, dy \tag{7-9}$$

This integral is extended over the cross section S of the neck
in the plane Oxy. This formula does not suppose that the
neck has the shape of a cylinder of revolution. However, in
this case the integration cannot be carried out in closed
form and must be done numerically. It presents no particular problems.

7-4 Cooling of a lava flow

Observation shows that the surface of a lava flow sets
very quickly, and since lava is a poor thermal conductor, the
surface is approximately at ambient temperature, which we
will take as 0°C.

As long as part of the lava remains fluid, with a latent
heat of solidification L cal/g, the cooling of the base continues as we have calculated for a dike. For the surface, the
formulas must be slightly modified to account for the condition $\theta = 0$ for $x = 0$. The equations then become

$$\theta = \theta_2 \, \text{erf} \left(\frac{x}{\sqrt{4\alpha t}} \right) \qquad \theta_1 = \theta_2 \, \text{erf} \, (\lambda)$$

and

$$\frac{\bar{\omega}L}{C} = \frac{\theta_1 e^{-\lambda^2}}{\sqrt{\pi\lambda} \, \text{erf} \, (\lambda)} \tag{7-10}$$

(see Fig. 7-1). We will calculate λ by Eq. (7-10), then for θ as
a function of x and t,

$$\theta_2 = (\bar{\omega}L/C)\sqrt{\pi}\lambda e^{\lambda^2} \tag{7-11}$$

When the solidification from the base and the top, which progresses at different speeds, meets at a point whose position as a function of θ_1 and L can be calculated, these formulas will no longer be applicable, and we will consider as previously for the dike, an initial fictive temperature $\theta_2 = \theta_1 + \bar{\omega}L/C$, to study further cooling. To satisfy the condition $\theta = 0$ for $x = 0$, we will introduce an antisymmetric initial fictive profile with a temperature $-\theta_2$ for $-e < x < 0$. Then the temperature is given by

$$\theta = \tfrac{1}{2}\theta_2 \left[2 \operatorname{erf}\left(\frac{x}{\sqrt{4\alpha t}}\right) - \operatorname{erf}\left(\frac{x - e}{\sqrt{4\alpha t}}\right) - \operatorname{erf}\left(\frac{x + e}{\sqrt{4\alpha t}}\right) \right] \qquad (7\text{-}12)$$

In the immediate vicinity of a lava flow, the temperatures attained will be practically the same as for a dike. But at a distance greater than its thickness, the temperatures are much lower. The calculation for the maximum temperature attained would be a little more complicated, but could be done according to the same principle.

7-5 Mechanical stresses of thermal origin

We have just seen how it is possible to calculate the temperature evolution at a point situated in the vicinity of a lava flow or dike. The unequal distribution of temperatures at a given moment leads to thermal dilatations which are not compatible with the preexisting mechanical stresses and which will consist of an increase in the pressure parallel to a dike or a flow or tangentially around a neck (but then, there would also be an increase in the radial pressure).

During a relatively long heating period, it can happen that there is a certain creep which reduces these stresses and tends to establish a hydrostatic distribution. During subsequent cooling, relative tensions will appear parallel to the walls of a dike or tangentially around a neck. It would also be necessary to account for stresses outside of the dike.

The results depend too much on the assumed hypotheses—as much on the initial stresses as on their relaxation during the heating phase. These hypotheses are necessarily gratuitous so that it does not seem useful to present an example of such a calculation here. Nevertheless, such stresses may play a part in the development of columnar jointing.

7-6 Cooling of a batholith

Until now, we have considered dikes or injected bodies, which are relatively thin with respect to their extension, a fact which allowed us to neglect the influence of the earth's surface. On the other hand, the resulting formulas show that the cooling time increases as the square of the thickness. This leads us to seek the cooling conditions for a granitic batholith, which can be very large, and which is emplaced underground under conditions which are still poorly understood. It is interesting to try to specify the order of magnitude of the cooling time, and, to that end, we will consider a schematic model with the top of the batholith at a depth h with an indefinite thickness. This batholith will be quickly brought up to a temperature θ_1, greater than the temperature which would normally occur at the depth h. We will not include any heat of crystallization. The condition $\theta = 0$ at the surface, for $x = 0$, is satisfied by considering a negative temperature profile, antisymmetric with respect to the surface. It is immediately found that

$$\theta = \tfrac{1}{2}\theta_1\left[\mathrm{erf}\left(\frac{h + x}{\sqrt{4\alpha t}}\right) - \mathrm{erf}\left(\frac{h - x}{\sqrt{4\alpha t}}\right)\right] \tag{7-13}$$

The gradient at the surface is given by

$$\left(\frac{\partial\theta}{\partial x}\right)_0 = \frac{2\theta_1}{\sqrt{4\pi\alpha t}}\exp\frac{-h^2}{\sqrt{4\alpha t}}$$

It is easily verified that the gradient passes through a maximum for $t = h^2/4\alpha$. This maximum has a value $0.482\,\theta_1/h$, and if the time is expressed in years and h in kilometers, then Eq. (7-13) yields $t_{\mathrm{years}} = 63{,}500h^2$ km. Thus, if the depth $h = 5$ km, $t = 1.6 \times 10^6$ yr (see Fig. 7-3).

Before the gradient reaches its maximum value at the

surface, it has higher values underground and it can happen that the temperature is high enough so that convection comes into play with rising steam (analyzed in Chap. 4). It is equally possible that water released by the crystallization of granite (which was formerly in solution in the silicate magma) is added to this steam. But the heat transport by water which escapes the magma is only a small portion of the heat which is liberated by the magma's crystallization and cooling. The major portion must escape by conduction or secondary convection.

The total heat dissipated at the surface can be calculated, per square centimeter, by integrating the flux $Q = K \int \partial \theta / \partial x \, dt$, and it is found to be

$$Q = Ch\theta_1 \left[\frac{\sqrt{4\alpha t}}{h\sqrt{\pi}} e^{-h^2/4\alpha t} - \text{erfc} \left(\frac{h}{\sqrt{4\alpha t}} \right) \right]$$

$$= Ch\theta_1 \frac{\phi(z)}{z\sqrt{\pi}} \tag{7-14}$$

where $z = h/\sqrt{4\alpha t}$. At the time $t = h^2/4\alpha$ (maximum flux), the total heat released is $0.165 Ch\theta_1$. This is the total heat contained in a slab of granite with a thickness one-sixth of its depth. Of course, the cooling will continue afterward, but it will come from lower and lower levels in the granite, and because of that will slow down.

Figure 7-3 indicates the temperature profiles at various times and can be used for any values of h and θ_1.

These results allow us to estimate under what conditions the emplacement of a granitic batholith (at a depth of several thousand meters) could perhaps give rise to a geothermal site, according to the hypothesis proposed for Larderello.

7-7 Phreatic explosions

Sometimes volcanic eruptions are confused with explosions in which the ejected materials are exclusively preex-

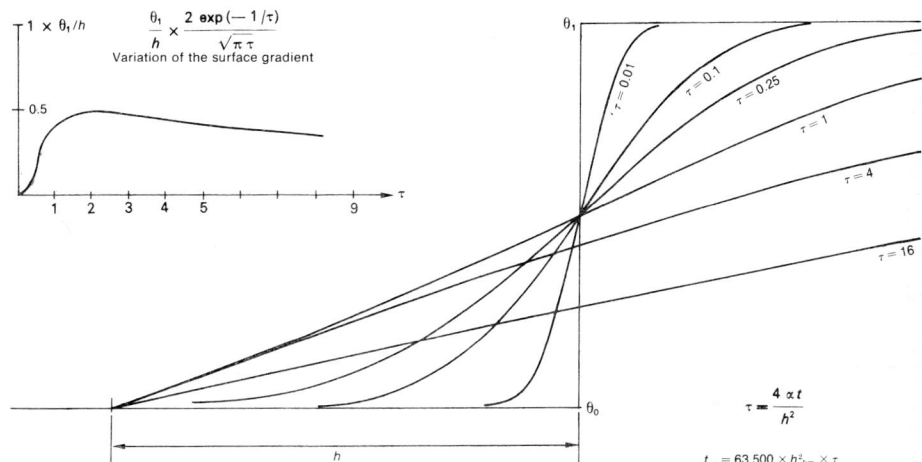

FIGURE 7-3   Thermal profiles for the cooling of a granitic batholith of infinite height; account is taken of the prior normal gradient and variations of the surface gradient.

isting rocks, without any lava, remaining at relatively low temperatures and sometimes damp. Such explosions, called phreatic, result from vaporization of water which impregnates porous rocks and becomes heated to a temperature greater than 100°C underground. In the expansion to atmospheric pressure, the water partly vaporizes while cooling, but the rocky matrix reheats this water, and if the water, steam, and rock remain intimately associated, which is plausible if it is a case of a massive pulverization, the ultimate result at atmospheric pressure can be either a suspension of rock in dry steam at a temperature greater than 100°C or a mixture of water and saturated steam in a proportion x, with the pulverized rock, at a temperature of 100°C. If $\bar{\omega}$ is the porosity (water mass per unit volume), 2.7 the specific gravity of the rock, and C, the heat capacity, is 0.5 cal/°C · cm³, then the final state after expansion and the mechanical work available in that expansion can be calculated[1] as a function of the initial temperature θ and the porosity, the deduction being made from the amount of mechanical work which is absorbed by atmospheric pres-

[1] Jean Goguel, "Le régime thermique de l'eau souterraine," *Ann. Mines,* Paris, X, p. 24, 1953.

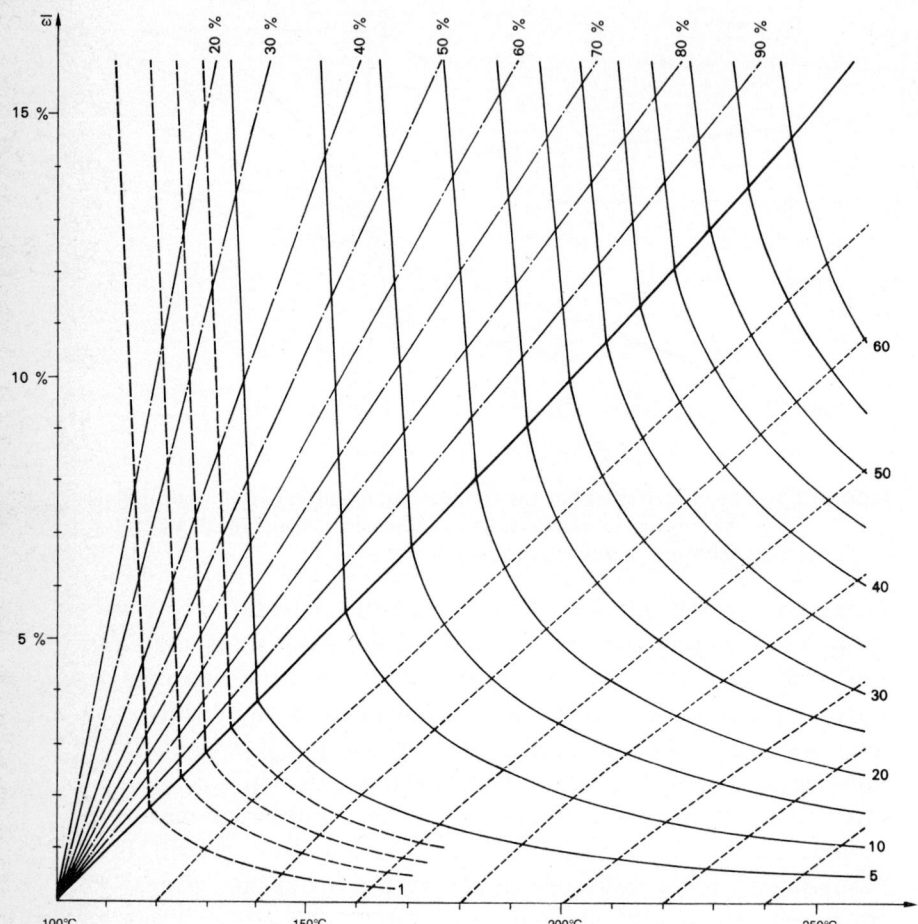

FIGURE 7-4   Graph giving the work available in an adiabatic expansion of a
humid soil up to atmospheric pressures as a function of the
porosity and the temperature (in joules per cubic centimeter).

sure. Figure 7-4 indicates the value of that energy in joules
per cubic centimeter. It is easy to see that the temperature
distributions whose origins we analyzed in Chap. 4 under
the influence of phreatic pressure represent a considerable
potential energy in such an expansion. The explosion could
be initiated by the opening of a fissure (possibly by a drill
hole) which would accidentally cause an eruption and thus
empty the water column which maintained pressure on the
fissure walls. The same energy release can be produced by a

depression of the phreatic layer which allows water to descend to contact with hotter rocks or by a lowering of the water pressure at constant temperature, which can also cause its vaporization. It is hardly possible to predict how the conclusion of this explosion will occur, since it must propagate in the walls of a funnel-shaped region while material is being released by the expansion. In any event, this explosion is not instantaneous since it implies the transport of heat by conduction from the centers of the rock grains to their surfaces.

Some historical examples of phreatic explosions are known, such as the explosion at Bandai-San, Japan, on July 15, 1888, or Usu-San, on the island of Hokkaido, in July 1944, and of Taal in the Philippines. Numerous craters (some filled with lakes) are found in New Zealand, which have been established to have resulted from phreatic explosions only a few centuries ago.

7-8 Heat produced by tectonic deformations

Heat is not the only form of internal energy. Geology has established that rocks have sometimes been profoundly deformed and folded by forces originating internally. Although these deformations are often very gradual, earthquakes can be imagined as being of analogous origin. Part of the earthquake energy propagates through the earth's crust in the form of seismic waves. Internal energy also manifests itself in upheavals which create relief, which erosion then modifies and finally destroys. These upheavals sometimes appear to us as a simple reestablishment of isostatic equilibrium. But in order for this equilibrium to have been destroyed, it would be necessary for subterranean density variations to be produced which absorbed mechanical work which was ultimately expressed in the surface upheaval. We will examine in the next chapter the manner in which we can interpret the mechanical energy thus released. Here, we will limit ourselves to evaluating the heat production which can result from such mechanical phenomena.

When a solid body such as a rock suffers a permanent deformation under the action of a stress to which it is subjected, the mechanical energy expended per unit volume is given by $\Sigma\sigma_{ij}\epsilon_{ij}$, where σ_{ij} represents a component of the stress tensor and ϵ_{ij} a component of the strain tensor. The best approximation that can be given for the law through which the deformation is produced is to suppose that the deformation occurs at a constant volume ($\Sigma\epsilon_{ii} = 0$) and that it is produced at the moment when the second invariant of the stress deviator reaches a limit S (which can be a function of the average pressure $\frac{1}{3}\Sigma\epsilon_{ii}$, but which is often treated as a constant). Let $S \leq D$ and $6D^2 = 2\Sigma\sigma_{ii}^2 - 2\Sigma\sigma_{ii}\sigma_{jj} + \Sigma\sigma_{ij}^2$, and let the components of the deformation be proportional to the components of the stress deviator. Then, it can be demonstrated that the work absorbed is given by $2S\sqrt{\Sigma\epsilon_{ij}^2 - \Sigma\epsilon_{ii}\epsilon_{jj}}$. For the most part, this work is transformed into heat. Only a small portion is found in the forms of defects of crystal dislocations.

The expression under the radical only depends on the geometrical deformation and in many cases its value can be estimated for a finite deformation, knowing the initial and final states and supposing that the passage from one state to the other occurs in the simplest fashion. For example, for a homogenous crushing which changes the height from h_1 to h_2, the energy absorbed per unit volume at constant thickness is $2S \ln (h_1/h_2)$, and $S\sqrt{3} \ln (h_1/h_2)$ if the crushing obeys rotational symmetry.

The study of tectonic deformations permits us to directly estimate the amplitude of the geometric deformation, which in a folded mountain range, such as the Jura (France-Switzerland), is on the average much less than one and generally only a few percent at most. If we knew the values for the threshold of plasticity, we could thus calculate the energy absorbed and almost completely transformed into heat. But it seems that there exist very slow deformation processes, by dissolution and recrystallization of matter, which play an essential role in geological deformations and which cannot be reproduced in the laboratory where rapid deformations are produced by totally different processes. It is clear that the strengths measured in the laboratory correspond to higher threshold values S than occur with slow geological deformations; we do not know in what ratio, but we know that laboratory experiments for threshold values furnish an upper limit for energy dissipation. If we assume that the compressive strength measured in the laboratory for the Jura mountains reaches a value of 1,000 bars (which is an overestimate for many of them), then $C = 1/\sqrt{3} \times 1,000$ bars. For a deformation of 1 percent, the absorbed energy is equivalent to 0.225 cal/cm^3, which raises the temperature 0.45°C.

Sometimes it happens that certain geological beds are much more deformed than others, for example, contacts along which the competent strata slide in the course of the folding. But if the folds are produced in this manner, it is because it would absorb less energy than a homogenous crushing affecting all the strata equally. The calculation of absorbed energy taking into account the entire deformation will thus give us a value that is too high. As folding is certainly a slow deformation, the heat released along certain strata can diffuse, and we can conclude that the temperature rise resulting from a folding such as that of the Jura or sub-Alpine ranges is at most a few degrees and probably much less. There can be zones where the general average deformation is much higher, but the temperature rise, even if it reaches one or several tens of degrees in places, is certainly much too low to be able to explain the metamorphism which is often associated with strongly deformed zones. The inverse relation can also be suggested, and it may be said that when the temperature is raised sufficiently so that new minerals are formed, or older minerals recrystallize, this recrystallization allows a slow rock deformation, even under the action of quite weak differential stresses, which could allow a large-amplitude deformation with only a moderate expenditure of energy.

7-9 Heat produced by faults

This situation is different only in the case of faults whose displacement can have a finite amplitude in a very short time, particularly at the time of earthquakes. The heat dissipated by friction along the surface only diffuses over a small distance and the temperature can increase appreciably. This phenomenon is much more important than in the preceding section, as this temperature rise can modify the frictional conditions and allow a much easier movement. Of course, it is only necessary to take into account the effective pressure, which is transmitted by the solid and which is less than the total pressure by the amount of the pressure of the fluid which saturates the rock. In certain cases, the fact that this fluid pressure is high can reduce the friction and facilitate the movement. Let p be the normal component of the pressure, f the coefficient of friction during the movement, which we will consider as constant, and v the speed of the movement. The power dissipated by friction per unit surface is pfv, and its equivalent heat q is dissipated on the two sides of the fault. The temperature distribution at a time t after the beginning of the movement (x being the distance to the fault) is given by

$$\theta = \frac{q}{C} \int_0^t \frac{1}{\sqrt{4\pi\alpha t}} \, e^{-x^2/4\alpha t} \, dt = \frac{qx}{2\sqrt{\pi}\alpha C} \, \Phi \, \frac{x}{\sqrt{4\alpha t}} \qquad (7\text{-}15)$$

In the plane of the fault, this reduces to $\theta = (q/C)\sqrt{t/\pi\alpha}$. It is easy to see how quickly this temperature increase occurs. Suppose that $p = 100$ bars (i.e., the hydrostatic pressure at 400 m), $f = \frac{1}{3}$, and the fault velocity is v cm/s. Then $q = 0.8v$ cal/cm$^2 \cdot$ s, which gives $\theta = 9v\sqrt{t_s}$. This temperature can also be expressed as a function of the displacement $h_m = vt$ and the duration of the movement $\theta = 9h_m/\sqrt{t}$.

When large earthquakes occur, fault displacements are of the order of one to several meters and the movement occurs in a fraction of a second. The formula above would indicate a temperature of several thousand degrees, but such a temperature would obviously modify the frictional conditions. This must have something to do with the formation of slickensides, which are sometimes observed in a displacement of only a few centimeters and which must lead to very low friction. But especially, this heating can increase the fluid pressure and this reduces the effective pressure transmitted by the solid, which alone produces the friction.

It can happen that hydrated minerals, such as gypsum, serpentine, and perhaps goethite (hydrated iron oxide) may exist along the fault. Once the dissociation temperature is reached (105°C for gypsum, 450 to 500°C for serpentine, and 150 to 170°C for goethite), the mineral decomposes and the water liberated contributes to the increase in the local fluid pressure. This effect can make that pressure equal to the total mechanical pressure and suppress all friction, allowing the displacement to continue without any resistance.

If the total pressure is less than the critical pressure of water, the heating in the plane of the fault may vaporize the groundwater and the pressure of the vapor may reach the value of the total mechanical pressure, which will also lead to a suppression of all frictional resistance.

But if such an event occurred, heat diffusion would immediately begin to lower the temperature, leading to a drop in fluid pressure, and thus the friction would reappear. Disregarding the fact that the system can be locally unstable, with accidental pressure drops succeeded by periods of nearly no friction, an ideal state can be conceived whereby there would exist only enough friction to maintain the temperature in the plane of the fault at a value for which the fluid pressure equaled the mechanical pressure. It does not matter if this is continuous friction or a series of shocks where friction occurs briefly between opposing rough areas in the fault surface. The heat flow in calories per square centimeter per second necessary for this case will vary as a function of time. It ought to be such that

$$\theta_s = \frac{q_0}{C} \int_0^{t_1} \frac{1}{\sqrt{4\pi\alpha(t-\tau)}}\, e^{-x^2/4\alpha(t-\tau)}\, d\tau$$
$$+ \int_{t_1}^{t} \frac{q(t)}{\sqrt{4\pi\alpha t}}\, e^{-x^2/4\alpha t}\, dt$$

knowing that

$$\theta_s = \frac{q_0}{C} \int_0^{t_1} \frac{1}{\sqrt{4\pi\alpha t}} \, e^{-x^2/4\alpha t} \, dt \qquad (7\text{-}16)$$

The approximations which were made do not justify the effort necessary to solve this equation, and we can be satisfied with a limiting solution $q(t) = q_0 \sqrt{t_1}/(\sqrt{t} + \sqrt{t - t_1})$. This expression shows that $q(t)$ will finally decrease as $1/2 \sqrt{t}$.

Table 7-1

$t = t_1 \times$	1.5	2	5	10	100
$q = q_0 \times$	0.585	0.414	0.235	0.161	0.05

But it can equally well happen that the fluid released, the presence of which lessens the friction, can escape, and it would require a greater release of heat than can be provided by friction to compensate for its loss. In particular, this situation can occur when friction vaporizes groundwater. The steam can escape when the plane of slippage reaches the surface, which is necessarily the case for a landslide, and several examples are known in which this process has played a role.[1] But this leakage which occurs at the periphery of the fault will play a relatively less important role when the sliding surface is more extensive, and it appears that when this sliding surface exceeds 1 km^2, the steam leakage is unimportant.

But the steam can also infiltrate the surroundings of the fault plane, compressing water before it, if the rock is sufficiently permeable. To analyze this phenomenon, it is necessary to account for the apparent compressibility of water (taking into consideration the elasticity of the pores in the rock). If the mass of water per unit volume is written

[1] J. Goguel and A. Pachoud, "Géologie et dynamique de l'écroulement du Mont Granier, dans le Massif de Chartreuse en Novembre 1248." *Bull. BRGM* (2d série), Sec. III, no. 1, pp. 29–38, 1972.
J. Goguel, Why Are Large Landslides Different from Small Ones?, in Barry Voight (ed.), "Geology and Mechanics of Rockslides and Avalanches," Pennsylvania State University Press.

as $\bar{\omega}(1 + \lambda P)$, Darcy's law allows one to write

$$\mu \nabla^2 P = \omega \lambda \, \frac{\partial P}{\partial t} \tag{7-17}$$

where μ is the permeability. Equation (7-17) is identical in form to the heat equation. The fluid pressure, which assumes large values in the plane of the fault, is thus propagated in the ground in the same manner as the temperature. If the propagation velocity of the fluid pressure is more rapid than that of the temperature in a permeable rock, the steam driving the water would make contact with cooler rocks where it would condense, and it would require a large increase in temperature along the plane of the fault to compensate for this condensation. Conversely, if the temperature diffuses more rapidly than the fluid pressure, the vaporization will progress evenly and the pressure can easily be maintained.

For an intermediate case, a more detailed description including the heat of vaporization can be undertaken to calculate the temperature—greater than that strictly necessary to guarantee vaporization under the existing pressure—which must be attained in the plane of the fault in order that the pressure is maintained in spite of steam penetration into the fault sides.[1] However, it will suffice to compare the equations for the propagation of the pressure and temperature in order to ascertain that if $(\mu/\bar{\omega}\lambda) < \alpha$ the pressure will be maintained, but not if $(\mu/\bar{\omega}\lambda) > \alpha$.

The apparent compressibility of water λ is of the order of $\lambda = 4$ to 7×10^{-5} bars^{-1} and $\mu/\bar{\omega} < 4$ to 7×10^{-7}, and depending on whether $\bar{\omega} = 1$ or 10 percent, this gives $\mu < 4$ to 7×10^{-5} to 10^{-6}. This permeability should be expressed in darcys if the water viscosity was 1 centipoise. At high temperatures, the water's viscosity decreases (it is four times less at 100°C and eight times less at 200°C) and the limiting

---

[1] Jean Goguel, Le rôle de l'eau et de la chaleur dans les phénomènes tectoniques, *Rev. Geogr. Phys. Geol. Dyn.*, vol. XI, no. 2, pp. 153–163, April-May 1969.

value of the permeability should be reduced in the same ratio.

The limit which has just been calculated should not be considered as a fixed limit. The processes described would certainly be able to act if the permeability was of an order of magnitude less than the limit considered above. In addition, it suffices that this value of the permeability be valid over a small thickness, since if the movement is rapid, the heat and pressure increase will only penetrate a small distance (several centimeters or several decimeters). An impermeable stratum in a lithologic series composed mostly of permeable strata can allow slippage according to the mechanism envisaged, if the slippage follows a stratigraphic contact. Near the limiting condition, it should be expected that even if the vapor pressure increases, it will only partially reduce the value of the effective pressure.

When the friction is reduced by such a process of vaporization, we have seen that, even though the velocity of motion is only several centimeters per second, the frictional resistance falls rapidly after a very small displacement. Thus, it is perfectly possible that the beginning of movement is not simultaneous along the entire plane of the fault. At the moment of rupture, the loss of the shear stress (which occurs because the moving friction is less than the stationary friction) leads to an elastic deformation of the two sides of the fault with the rupture propagating at a finite speed (but much greater than the relative velocity of the two sides). The beginning of the vaporization process can follow the propagation of the rupture with a slight lag, and so for the total resistance to gliding, which results from an average taken over the entire surface area, we will not have to consider successive phases as for the displacement of a specific surface element.

These phenomena can play an important role by allowing displacements of large dimensions and by facilitating overthrusting, which constitutes one of the structural characteristics of certain mountain ranges such as the Alps or Canadian Rockies. But their thermal role remains very limited. The ease of gliding greatly reduces the mechanical energy absorbed and thus the total heat produced. The temperature rise remains confined to the immediate neighborhood of the gliding surface, and as soon as the movement is terminated, this heat diffuses away and the temperature quickly returns to normal. The briefness of heating (the duration of which can be counted in minutes) explains why we generally do not observe metamorphism, which would allow us to estimate the maximum temperature attained by the presence of characteristic minerals.

7-10 Metamorphism

We have seen that the thermal equivalent of mechanical energy dissipated in tectonic deformations remains, on the average, very limited. The temperature rise demanded by metamorphism cannot be caused by these processes. On the contrary, we will see in the next chapter that the mechanical energy expended in these deformations must be considered as *derived* from the geothermal flux.

The essential cause of metamorphism must be considered to be elevated temperature (without excluding the possibility that this high temperature occurs only in rocks carried to great depths by tectonic deformations) at a given pressure, and the rock could include a fluid phase whose role can be important. Under these conditions, certain minerals are no longer stable and others can replace them. There is crystallization of new minerals or recrystallization of old ones.

This recrystallization, occurring in a solid body, is strongly influenced by the system of mechanical stresses which occur and which determine a preferred orientation for the crystals. This constitutes one of the most striking features of metamorphic rocks. At the same time, this recrystallization permits a relatively slow and easy deformation of the rock.

From a thermal point of view, we will not have to be concerned with the conditions that produced the temperature rise which is the origin of metamorphism and with the physical state of the fluids which can play a role in these processes.

The schematic model of a batholith studied previously and the heating which results in the country rock can give an idea of the necessary delay period for an appreciable temperature rise in a large volume of rock. Even if such a model does not seem to satisfactorily explain the temperature rise in metamorphism, it is actually not clearly understood how we could represent this temperature rise.

Mineral transformations, characteristic of metamorphism, imply a certain heat of reaction. As a general rule, a temperature increase must promote endothermic reactions and the heats of reaction can be several tens of calories per gram of new mineral. This can appreciably slow the heating. Guitard and Fonteilles remarked that when a sedimentary cover lay on an older basement which had experienced a prior metamorphism and whose constituent minerals were thus in a high temperature equilibrium, the heating of the basement would be much faster than the heating of the cover where the formation of new minerals absorbs heat. Thus, we should count on, if not a discontinuity in the maximum temperature attained (attested to by the metamorphism), at least a discontinuity in the gradient at that temperature. This is what has been called the *basement effect.*

7-11 Possible fluid phases

Certain secondary aspects of metamorphism, and in particular vein deposits, are often attributed to fluids circulating along fissures. To describe these fluids, the words *hydrothermal* and *pneumatolitic* are often used. This last word implies the action of a gas or a vapor. It is useful to try to specify the different fluid phases which may be involved.

We have previously described the way in which two phases, water and steam, are no longer distinguishable for pure water beyond the critical point (221 bars, 374.11°C). But the same question should also be examined for mixtures or solutions.

Carbon dioxide (for which we do not have to distinguish between liquid and gas, the critical point being very low), although it has a mutual solubility with liquid water, can constitute a distinct phase from water, while, like all gases, it is totally miscible with steam. The limit of existence of these two phases extends beyond the critical point of pure water, up to a temperature which slowly decreases as the pressure increases (350°C at 365 bars, 300°C at 600 bars). However, it is doubtful that the "wet CO_2 phase" plays an appreciable geological role.

In the presence of an excess of soluble salt, the salt contents of water and vapor phases in equilibrium with each other are generally very different. They are not equal at the critical point (as opposed to the properties of pure water), and above the critical point of pure water there can coexist two distinct fluid phases with very different salt contents. For sodium chloride, the coexistence domain between the two phases extends up to the fusion point of anhydrous salt. The pressure at which the two phases are in equilibrium is less than the vapor pressure of pure water, but the pressure continues to rise beyond the critical temperature of water until reaching 389 bars at 600°C. Then the pressure decreases and is equal to atmospheric pressure at 804°C, the fusion point of pure salt. The dense phase then consists of a molten salt whose fusion point is lowered by the presence of water in solution, and there is also a gaseous phase, a mixture of water vapor and sodium chloride vapor (whose vapor pressure is small). At a sufficient temperature, there is thus a continuous passage from the aqueous salt solution to the molten salt.

For other less soluble substances—particularly for silica—there exists a critical point for saturated water in equilibrium with the solid phase. This critical point is slightly displaced with respect to the critical point of pure water. Nevertheless, the laws of gas mixtures (addition of partial pressures) are totally inapplicable and the solubility of silica in supercritical water is relatively high.

At the fusion temperature of silicate solutions, water can be dissolved in the melt with a percentage (on a weight basis) which increases much less quickly than the vapor pressure of water, and which can go to 3 percent at 500 bars and 9 percent at 4,000 bars. Conforming to the law of gas solubility, this percentage decreases slightly when the temperature increases. In addition, it is known that the presence of this water greatly lowers the solidification temperature of silicates (from 100 to 200°C). But if the silicates crystallize, it is in an anhydrous form and water must be released. Thus, again we are concerned with two phases, one of which is identified with the mixture of molten silicates and the other with supercritical water, saturated with respect to the silica and other oxides which are present. This aqueous phase is liberated at the time of silicate solidification, and if it escapes along a fissure, its temperature and pressure will vary and thus it can release the oxides which it held in solution. Such is the origin of hydrothermal veins. It is only near the surface, depending on the temperature distribution, that supercritical water will reach either the liquid water domain, where the solubility is high for many substances and where mineralization can occur to form an ore deposit, or the steam domain, an exceptionally hot region, making a fumarole. Well before reaching the surface this will have essentially no solids in solution and will contain practically no impurities except other gases.

Thus, it does not seem necessary to distinguish between pneumatolitic and hydrothermal deposits. There can be a difference between the deposits formed at the expense of the volatile phase in the vicinity of a magma in the course of crystallization (perhaps certain pegmatites) and those deposits left by thermal waters at moderate depth, perhaps with differences caused by the temperature and the pressure. But it does not seem that there can be a fixed distinction between the two discontinuous fluid phases.

It would theoretically be different in the case where the solid constituents were soluble enough to act like sodium chloride; the water vapor phase would be able, in the path by which it escaped to the surface, to meet the saturation curve and pass from the vapor state, with a very low concentration of dissolved salt, to the liquid state with a capacity for considerable dissolved salt. This passage is what has been called *regressive boiling* (i.e., in fact, a condensation). It would be marked by a very strong dissolution of previously deposited salts, appearing as corrosion. But it does not seem that this phenomenon, which has been theorized, plays a very important role.

The detailed study of the solubility and stability of different minerals and the conditions in which they can be deposited goes beyond the scope of this book and will not be further discussed.

geothermics on a global scale

As soon as the existence of a gradient and thus a geothermal flux was discovered, the question was asked about the origins of the heat so dissipated. In the eighteenth century, the answer seemed evident to Buffon: the earth was cooling off, and he tried to estimate the earth's age according to these cooling conditions. Not possessing any valid theory of heat, he measured the cooling times for musket balls and cannonballs of various diameters (from ½ to 6 in.), heated in his kilns at Montbard. These times being proportional to the diameters, he unhesitatingly extrapolated them to the dimensions of the earth and got an age of 63,000 yr (according to his manuscripts, Buffon hesitated between this figure and an age ten times greater). One of his errors was to consider a ball as cooled when it could be "held in the hand without the hand being burnt." This obviously corresponds to a thermal gradient much higher than that of the earth.

The calculation of Buffon was performed again several times during the nineteenth century. It inspired Fourier to construct his theory of heat and was finally brought up to date by Lord Kelvin.

In the meantime, Helmholtz's attention had been drawn to another source of energy, which is essential for the stars, particularly the sun but which is not negligible for the earth: in the contraction of a celestial body under its own gravitational attraction, from what may be taken to be a cloud of matter initially dispersed over large distances, gravitational potential energy is dissipated and the greater part of that energy is transformed into heat. This energy suffices to explain how the stars were brought to very high temperatures, and it would have supplied enough solar radiation for 10 million yr, which led Lord Kelvin to give this as a limit to the age of the solar system, but not without incurring some objections from geologists.

It is well known that the discovery of radioactivity and then nuclear fusion completely changed the perspective of this problem. Even though the earth's rocks are low in radioactive elements, their radioactivity is such that the heat produced is of the same order of magnitude as the geothermal flux and thus constitutes an essential factor in the thermal history of the earth. On the other hand, nuclear fusion reactions, such as occur in the transformation of hydrogen to helium, have been able to supply the solar radiation for 4.6 billion yr, which is the age attributed to the solar system, and will continue to do so for an even greater length of time.

In a global thermal balance, it can be shown that possible chemical reactions do not play an appreciable role. On the contrary, tectonic activity, by which rocks are deformed, mountain ranges raised, and energy dissipated in earthquakes and perhaps in other forms in relative plate displacements, represents a mechanical energy which must be accounted for in a total energy balance. Again it can be shown that this term is very small compared to the geothermal flux; it can thus be imagined that this mechanical energy can be derived, by a mechanism to be specified, by conversion from thermal energy but not the inverse.

In the energy balance, the magnetic field, or rather the energy dissipated in the course of its modifications, represents a much less important term. But its great significance is to give us information about the regime existing in the earth's core, of which we possess little knowledge.

Thus, the ideas of the thermal history of the earth for two centuries have evolved in a quite radical manner. Even now, the temperature distribution in the internal regions of the earth is certainly the most poorly understood physical property of the earth, particularly because the study of that distribution is inseparable from hypotheses of the earth's origin. It is hardly necessary to state how uncertain these hypotheses are.

We are going to undertake in more detail the study of the different terms in an energy balance just sketched to show how their orders of magnitude can be estimated. Afterward, entering the domain of hypotheses, we will try to show how the thermal system of the earth as a whole can be imagined and connected to its origins.

8-1 Cooling of a solid sphere

Obviously it is not very satisfactory to compare, as did Buffon, the cooling of the earth to the cooling of a cannon-ball (i.e., treating the earth as a homogenous sphere in which the thermal parameters such as the specific heat and thermal conductivity are considered constant). Nonetheless, it will be useful to recall the results of this model.

The heat equation for a sphere, when the temperature distribution is a function of the radius only, reduces to

$$\frac{\partial(r\theta)}{\partial t} = \alpha \, \frac{\partial^2(r\theta)}{\partial r^2} \tag{8-1}$$

This equation permits solutions of the form

$$\theta = A \, [\sin \delta r / r] e^{-\alpha \delta^2 t}$$

The temperature distribution of a sphere with an initial uniform temperature θ_1 at $t = 0$, and whose surface $r = R$ is afterward maintained at zero, is obtained by letting $\delta = n\pi/R$ and representing the initial temperature by a sum of the type

$$\theta = \frac{-2R\theta_1}{\pi r} \, \Sigma \left[\frac{(-1)^n}{n} \right] \sin \left(\frac{n\pi r}{R} \right) e^{-\alpha n^2 \pi^2 t / R^2}$$

(which is equal to θ_1 at $t = 0$). This equation gives a surface gradient of

$$\frac{\partial \theta}{\partial r} = \frac{-2\theta_1}{R} \, \Sigma e^{-\alpha n^2 \pi^2 t / R^2}$$

An approximation to the right-hand side of this equation can be obtained by replacing the summation by an integration, and thus we find that approximately:

$$\Sigma e^{-\alpha n^2 \pi^2 t / R^2} = \frac{R}{2\sqrt{\pi \alpha t}} - \frac{1}{2}$$

This allows us to calculate the value of the surface gradient:

$$\frac{\partial \theta}{\partial r} = \frac{\theta_1}{\sqrt{\pi \alpha t}} - \frac{\theta_1}{R} \tag{8-2}$$

The first term represents the gradient for a cooling of a half-space (infinite radius), and the second term takes the spherical shape into account.

On the whole, the successive terms in the series representing the temperature die out more and more rapidly, and at the end of enough time, only the first term remains appreciable. But for a sphere having the thermal properties of superficial rocks ($\alpha = 0.01$) and the earth's radius, this would be so only after a time on the order of hundreds of billions of years. For the age of the earth, 4.6 billion years, the initial temperature would barely have changed and the surface cooling would only have penetrated to a depth shallow compared to the earth's radius. A gradient of the order of 1°C/30 m would require an initial uniform temperature on the order of 25,000°C. Obviously, it would be absurd to suppose that at such temperatures a solid state exists with thermal properties analogous to those of superficial rocks.

This was essentially the model by which Buffon (in an incorrect manner because he misunderstood the significance of the gradient) and other physicists reasoned till Lord Kelvin. Lord Kelvin would still have found acceptable central temperatures for durations much greater than 10 million yr, which was the time to which he intended to limit the age of the earth (and for which $\theta_1 = 1000$°C) by his reasoning concerning the solar radiation.

For any other initial temperature distributions, an analogous conclusion would be reached. The simplest way to show this is to consider this temperature as the sum of a uniform temperature equal to the initial temperature at the surface and an initial temperature distribution canceling at the surface. A calculation similar to the preceding one would show that this second distribution generally brings only a negligible contribution to the gradient and would only have evolved slightly in a time equal to the age of the earth. The initial distribution of temperatures in the central regions of the earth has practically no influence on the surface gradient, which depends on the cooling conditions near the surface only. In the model of a solid sphere, the earth is much too young to allow the initial thermal conditions in the central regions to be manifested in any way at the surface. Of course it is extremely unlikely that this model corresponds to reality; it only constitutes a reference which it was necessary to recall.

8-2 The role of radioactivity

Since the discovery of radioactivity in 1896, the energy dissipation by radioactive elements has led to the reexamination of the whole problem of the thermal history of the earth. This is because the radioactive substances existing in the interior of the earth release heat at a rate which could be comparable in order of magnitude to the heat dissipated by conduction to the surface. To be more precise, it would be necessary to be able to estimate the amount of radioactivity found in the interior of the earth. First there is uranium and thorium. Natural uranium is composed of two isotopes; ^{238}U constitutes $\frac{137}{138}$ of the total uranium content. ^{235}U is less abundant but has a higher disintegration constant. This constant λ characterizes the disintegration probability in unit time such that the amount of radioactivity present varies as $e^{-\lambda t}$ and, in particular, is reduced to one-half its initial value in a time $t = 0.6932/\lambda$, often called the *half-life*.

Table 8-1 indicates, besides the values of λ, the heat release from 1 g of the substance supposed in equilibrium with all the products resulting from its disintegration.

Outside of uranium and thorium, we could also take into account potassium, whose isotope ^{40}K is scarce (0.011 percent) and which is radioactive through two different processes, producing argon and calcium, respectively. We need not distinguish between these processes in our study.

The effect of Rb, whose isotope ^{87}Rb makes up 27.8 percent of the total Rb content and which has a disintegration constant of 1.47×10^{-11} yr, is often neglected. Estimates of the abundance of ^{87}Rb give the same approximate figure as for the amount of ^{40}K. The extreme slowness of disintegration justified neglect of its heat production. The effect of rhenium, which is very scarce, can certainly be ignored.

Of course we do not have to include the radioactive element produced in the upper atmosphere by cosmic radiation (^{14}C, tritium, etc.). Their disintegration at the surface simply returns a portion of the incident radiant energy.

However, we will raise a very obscure point here: it is known that the neutrino is a particle whose existence has been established by theory to ensure the principles of conservation of mass and energy in certain reactions, but it has practically no interaction with matter and cannot be detected. It is sometimes asked if the neutrino flux received from the sun is such that the feeble absorption which would occur when it passes through the earth would lead to a heat release which should be taken into consideration. This would mean that should we admit the principle of energy conservation is faulty at a certain level, which is exactly the level in which we are interested. Some theoretical discussion, into which we will not enter, seems to show that even on this scale the effect would be completely negligible.

Table 8-1

	λ, yr^{-1}	cal/g $\cdot$ yr
^{235}U	9.72×10^{-10}	4.3
^{238}U	1.54×10^{-10}	0.71
U (ordinary)		0.73
Th	4.99×10^{-11}	0.20
^{40}K	$5.3 \ \times 10^{-10}$	0.21
K (ordinary)		27×10^{-6}

Once in possession of these facts, which can be considered as well established, it remains to estimate the quantities of radioactive elements present, and this is extremely difficult and uncertain. It is certain that the rocks accessible at the surface are only representative of a thin layer (a few tens of kilometers) which makes up the crust, and again, in proportions which ought to be weighted.

Samples of rocks which we can consider to belong in the mantle have very weak radioactivities (at the limit of what is measurable), and even then we may wonder if the samples have not been contaminated during the processes which brought them to the surface.

Meteorites, in particular chondrites, which are habitually considered as being representative of the interior of planets analogous to the earth, also have very small percentages of U, Th, and potassium. Clearly, the percentages of radioactive elements in the crust are much higher; but, even the small percentages in the mantle and core may produce an amount of heat which may not be negligible.

In the crust, there exist accumulations of certain elements which constitute exploitable deposits. Clearly, for the three elements in which we are interested, they only represent an infinitesimal proportion of the total existing tonnage and it can be shown that their thermal effect can be neglected. An occurrence of 100,000 tons of uranium would be quite extraordinary; at a content of 1 percent, that would occupy a sphere with a 100-m radius; the heat liberation of 2300 cal/s, supposed uniformly distributed in the sphere, would lead to a steady-state thermal system with a temperature increase of 3.6°C at the surface and 5.7°C at the center. Thus, this would only be a very local perturbation.

For many rocks, a nice correlation is observed (which is poorly explained but is perhaps not a coincidence) between the percentages of uranium, thorium, and potassium. For basic and intermediate rocks, the percentages can vary between 0.1 and 3 ppm for uranium, with an average of 0.9 ppm, and between 0.3 and 10 ppm for thorium, with an average of 2.7 ppm. On the other hand, for granites (in the broad sense), U percentages are found from 1 to 20 ppm and more, with an average of 4.9 ppm, and for thorium, from 2 to 20 ppm, with an average of 9.3 ppm. For these rocks, the potassium contents are from 3 to 4 percent.

The misleading significance of averages should be pointed out: they represent a weighting between samples and not between the tonnages they represent. Samples are more often collected at the edges of massive granitic rocks than in the centers. Radioactive elements (other than potassium) are, in granite, almost exclusively contained in certain accessory minerals such as zircon.

If certain sedimentary rocks, such as biogenic limestones where the radioactive contents are small or evaporites where they are practically nonexistent, are set aside, the proportion of U in sedimentary rocks varies with the proportion of clay, quartz, or heavy minerals. It appears to oscillate around an average of 2 to 3 ppm; the Th percentages have a less strict correlation than for eruptive rocks.

Certainly the "granitic layer" of seismologists should not be identified with the intrusive granites observed on the surface and which result from refusion processes probably leading to a concentration of Th and U. Gneiss on the other hand should retain percentages of U and Th much the same as those in the rock from which it originates.

In summary, it is very difficult to give numbers which can be considered as representative values for the average radioactivity of the crust. The act of taking for the examples average concentrations of 2 ppm for U, 8 ppm for Th, and 2 percent for K should not make us forget that these numbers can be off by a factor of 2 in either direction. These values lead to a heat release of 3.6×10^{-6} cal/g $\cdot$ yr, or 9.6×10^{-6} cal/cm^3 $\cdot$ yr; a thin layer of 39 km could furnish the average geothermal flux of 1.2 μcal/cm^2 $\cdot$ s. This thickness would be a little greater than the depth which is normally assigned to the continental crust, but we should remember that the orders of magnitude are similar.

In oceanic regions, on the contrary, what is known of the crust's composition (thinner and nongranitic) seems to indicate a heat production much lower than the geothermal flux, the average of which is the same as for the continents.

To estimate the significance of these terms, we can calculate the temperature rise if the radioactive rock was isolated. It is easily found, taking into account the law of variation of the quantity present as a function of time, that the temperature of this rock would have been raised 5770°C in the last 1 billion yr and 47,575°C during the last 4.5 billion yr. It is quite evident that conductive cooling in the crust is very important on this time scale. But, if we supposed that in the central regions of the earth the percentages were a specified fraction of those taken for the crust, we would be able to deduce from these numbers the temperature rise which, if only solid conduction had occurred, would only have been slightly modified.

It is indispensable when the thermal evolution of the earth is considered to account for the decrease in radioactive elements with time, which results in heat liberation being more rapid in the past. Figure 8-1 indicates for past ages the heat released for a composition such as we have previously imagined; it is seen that 3 billion yr ago the role of potassium in the heat release was preponderant and 4.5 billion yr ago it was the heat release by ^{235}U which was important (at that time, the percentage of ^{235}U was 27 percent of the total uranium content).

Observe that if no other indications of the earth's age existed, the extrapolation of these figures would lead us to fix an upper limit for this age. From Fig. 8-1, we can also deduce the variations of the geothermal flux during past geological times caused by the radioactive elements then contained in the crust. But for the portion which would arise from cooling from an initial state at a high temperature, it could well be that this cooling rate was higher in the far distant past.

The comparison of the flux value 4 billion yr ago (3.5 times today's flux) with the limit which we have shown in Chap. 4 for the hypothesis of a thermal stratification as it results from heat transfer by conduction would not be compatible with the presence of liquid water in a permeable terrain. This leads us to think that the formation of steam plumes and heat transfers by convection must have been

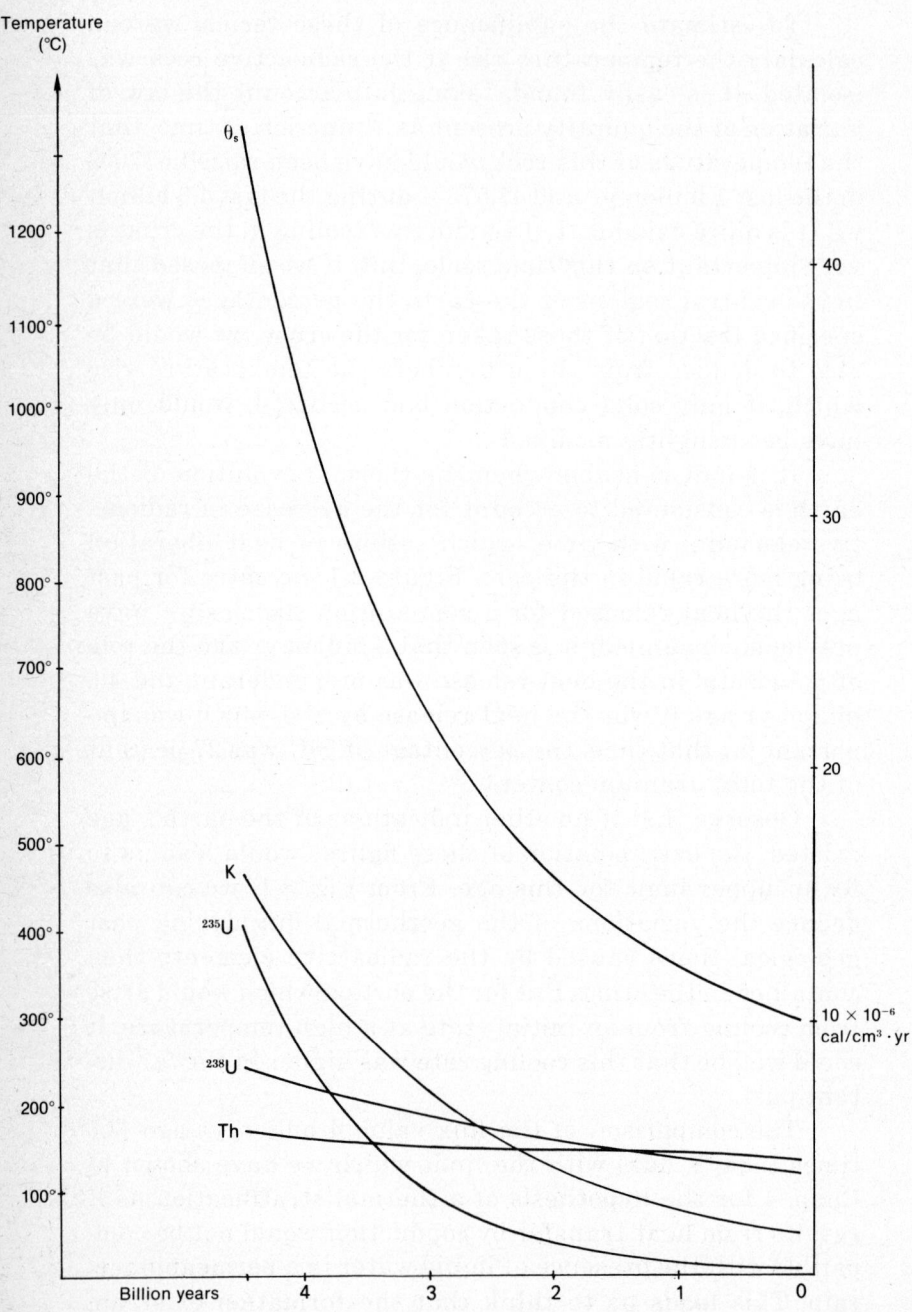

much more frequent than today; generally, all the subterranean processes in the crust must have been much more active. This is a fact which must be considered in the study of the lower Precambrian.

8-3 The cooling of the earth including radioactive effects

If we suppose that all radioactive substances are contained in the crust, one way of studying their influence on the cooling of the earth is by considering the temperature as the sum of two terms. One term (θ_1) satisfies the heat equation without taking into consideration the heat released by radioactive substances, and the other (θ_2) is taken to be zero below the base of the crust and takes into account the radioactive heat release, which is of the form $\Sigma q(x)e^{-\lambda t}$. For each of the four terms of this sum, let $\theta_2 = F(x)e^{-\lambda t}$, where $F(x)$ satisfies the equation

$$-C\lambda F(x) = K \frac{\partial^2 F}{\partial x^2} + q(x) \tag{8-3}$$

If the first term is neglected, we have $F(x) = (-1/K) \int_x^h q(y) \, (y - x) \, dy$, and at the surface $F(0) = (-1/K) \int_0^h q(x) \, x \, dx$. Supposing that the radioactive elements are uniformly distributed over a thickness of 30 km, it is then found that $\Sigma F(0) = 290°C$.

It is easy to see that the left-hand side of the equation is of the order of a hundred times less than each of the two terms of the right-hand side, which allows us to limit our-

FIGURE 8-1   Graph giving the heat release per cubic centimeter in 10^{-6} cal/yr, for a typical rock of density 2.8 and composed of 2 ppm of U, 8 ppm of Th, and 2 percent of K. The graph is given as a function of geological time; the amount of heat resulting from each element or isotope is shown. The superficial fictive temperature θ_s has also been indicated, and so its effect can be included in the study of deep subterranean cooling with the hypothesis of a uniform distribution of such a rock over a 30-km depth.

selves to the approximation furnished by the solution: $\theta_s = \Sigma \theta_i e^{-\lambda t}$, with $\Sigma \theta_i = 290°C$. In Fig. 8-1 the fictive temperature θ_s is indicated.

For θ_1, which satisfies the heat equation not taking the radioactivity of the crust into account, the condition $\theta = 0$ at the surface must be replaced by the condition $\theta = \theta_s$. Thus everything happens for θ_1, as if, instead of cooling through a surface kept at a constant temperature that is taken as 0°C and determined by equilibrium with the external radiation, this surface has been kept at a fictive temperature θ_s which progressively decreases from 1300 to 290°C.

In a manner analogous to what was just done, if an initial temperature distribution was given, it would be broken down into three parts: first, a uniform distribution equal to the difference between the initial surface temperature and 1300°C and which will evolve as we have seen above; next, an initial temperature distribution of 1300°C with a superficial fictive temperature progressively decreasing from 1300 to 290°C, according to the law indicated; and finally, an initial distribution zero at the surface and for which the surface temperature will remain 0°C. It has already been stated that this last portion would only bring a small contribution to the surface gradient.

Essentially we have to consider the second distribution to see what gradient results from a progressive lowering of the surface temperature. The calculation can be done for a sphere,[1] but it leads to a slowly converging series whose terms are exponentials and it does not seem useful to develop this calculation here. We will study the problem in the plane approximation.

It has been seen that if at the instant t the temperature at the surface of a half-space is lowered by $\Delta\theta$, the subsequent temperature will be given at time τ by

$$\theta = \Delta\theta \text{ erfc} \left(\frac{x}{\sqrt{4\alpha(\tau - t)}} \right) \tag{8-4}$$

[1] J. Goguel, Note sur le refroidissement du Globe, *Ann. Geophys.*, vol. IV, no. 3, pp. 253–258, 1948. At the time, we adopted slightly different thermal characteristics.

If the surface temperature, initially 0°C, is afterward given by $\theta_s = f(t)$, we will have

$$\theta(x,\tau) = \int_{-\infty}^{\tau} f'(t) \, \text{erfc} \left(\frac{x}{\sqrt{4\alpha(\tau - t)}} \right) dt$$

and

$$\frac{\partial \theta}{\partial x} = - \int_{-\infty}^{\tau} f'(t) e^{-x^2/4\alpha(\tau - t)} \frac{1}{\sqrt{4\alpha(\tau - t)}} \, dt$$

where for $x = 0$ and $\tau = 0$, this expression becomes

$$\frac{\partial \theta}{\partial x} = - \int_{-\infty}^{0} \frac{2f'(t)}{\sqrt{-4\pi\alpha t}} \, dt \tag{8-5}$$

The integral has been numerically calculated, supposing concentrations of the type indicated earlier distributed uniformly over a 30-km thickness. With $f(t) = \Sigma q e^{-\lambda t}$, which gives $f'(t) = \Sigma q \lambda e^{-\lambda t}$, the gradient is found to be equal to 1°C/480 m.

This shows that, because of the temperature drop at the surface, a thermal flux issues from the earth's interior with corresponding cooling and is dissipated at the earth's surface at the value of 0.1 μcal/cm$^2 \cdot$ s, which is approximately a tenth of the geothermal flux.

In fact, it is extremely doubtful if the interior of the earth can be assigned the thermal properties of superficial rocks, even if only the thermal properties of the upper part of the mantle were important, as would be the case in the schematic problem which has just been treated. In particular, it can happen that in the mantle, heat-transfer processes take place which are much more effective than conduction. We will see later that is is undoubtedly necessary to call upon convection. For the moment, we will limit ourselves to representing these processes by the choice of a higher value for the coefficient K, and thus also for α.

In the expression for the flux, K occurs in the numerator and the square root of α in the denominator under the summation sign. The effect of an increase in K is an increase in the thermal flux reaching the surface (or the base of the crust), as a consequence of the decline of the fictive temperature at the surface, which goes as the square root of the conductivity. Whenever convection plays an important role in thermal exchanges at the level of the upper mantle, it happens that in the geothermal flux, the portion resulting from the cooling imposed on the lower levels by the thermal evolution of the crust, represents a substantial proportion of the total thermal flow.

It is perhaps useful to emphasize that this effect does not depend on the total amount of radioactive substances present, but on an integral in which their depth occurs (as a moment of the radioactive substance's amount with respect to the surface). If it is supposed that in oceanic regions the processes which led to the concentration of uranium and thorium in the granitic rocks of the crust did not act in the same way as for the continents, it could be imagined that, even if the total radioactive concentration is lower than for the continents, this moment could be higher and thus play an appreciable role in the flux originating from underground caused by cooling, which could explain the equality of the average oceanic and continental fluxes.

Certain of our hypotheses are very arbitrary, in particular those which concern the amounts of radioactive materials and their distribution with depth in the crust. But we have the definite result that we do not have the right to imagine an earth whose geothermal flux would exactly be balanced by the heat released by radioactivity in the crust: from the sole fact of the decay of radioactive substances, there results a fictive temperature decline (and thus also for the temperature at the base of the crust), which leads to a flux caused by cooling, at least of the order of one-tenth the normal heat flow (and perhaps higher).

In the calculation of the integral which yielded this result, and in spite of the factor $1/\sqrt{t}$, the earliest geological periods play a preponderant role. In fact, this only holds for one of the initial temperature distributions, that part which corresponds to an initial uniform temperature equal to the fictive temperature due to the radioactivity of the crust, which therefore was not initially subjected to any tendency to cool. It would equally well be necessary to con-

sider the other parts of the initial distribution, one nonuni-
form and the other presenting an initial discrepancy with
the surface temperature. These two terms could easily give
contributions of the same magnitude to the actual surface
gradient.

As a case in point, although it does not seem at all
probable, let us suppose an initial temperature of 0°C everywhere and a radioactivity concentrated only in the
crust; we should add to the solution which has just been
studied, and which corresponds to an initial temperature of
1300°C, a solution corresponding to an initial uniform temperature of −1300°C. The calculation in Eq. (8-2) furnishes
us with a value for the corresponding gradient of 1°C/500
m, nearly equal and opposite to the gradient resulting from
the decay of radioactive substances. This shows that, first,
the radioactivity of the crust heats the underlying rocks
and, then, the crust begins to cool and a small amount of
the heat which was accumulated below returns to the
crust. There is no interest in a detailed investigation of a
model so far removed from reality, and we only mention it
as an example of the way in which a temperature distribution can be obtained which is appropriate to any sort of
hypothesis relating to the initial state.

If we knew the distribution of the radioactive elements,
it would be easy to calculate its consequences for the temperature distribution; unfortunately we lack precise information on the distribution of radioactive materials in the
mantle and especially the core. Comparison with meteorites,
particularly chondrites, furnishes the only model which we
can apply, but it is not known to what extent it is valid; it
can only be remarked that it furnishes us a strong justification for the notion of mantle concentrations much lower
than in the crust, but not zero. This confirms the conclusions
of a purely thermal analysis which shows that the heat
flow would be much higher for mantle concentrations
of the order of those in the crust, which can thus be formally
excluded. On the other hand, the necessity for an engine for
the internal mechanical activity, the consequences of which
are so visible on the surface, makes it likely that there is a

nonzero concentration in the mantle and in the core. Our uncertainty is perhaps even greater for the distribution in the oceanic regions.

8-4 The dissipation of gravitational energy in condensation

Helmholtz's attention was drawn to the energy source which is liberated when a mass, initially dispersed at a great distance, condenses under the effect of the mutual gravitational attraction of its parts and whenever the radial mass distribution is altered (we will assume that the spherical symmetry is preserved, and we will neglect the effects due to rotation). Part of this energy will be found in the form of elastic energy in the compressed matter, but the rest must be dissipated in the form of heat.

To calculate the energy liberated between dispersion at a great distance and a condensed state, we can proceed in the following manner. Consider the mass $m(r)$ interior to the radius r, such that $m = 4\pi \int_0^r \rho r^2 \, dr$. This mass exerts on a mass dm, between the radii r and $r + dr$, a gravitational attraction $\gamma = f(m \cdot dm)/r^2$, and if this mass was dispersed to infinity, the work done would be

$$dU = f\left(\frac{m \cdot dm}{r}\right) \tag{8-6}$$

For the entire sphere, the work done by dispersion to infinity would be equal to $U = f \int_0^m (m \cdot dm)/r$, if we consider r as a function of m, or if we consider m as a function of r, then $U = f \int_0^R m \, dm/dr \, (dr/r)$.

For a sphere with uniform density equal to 5.51, we have $m = \frac{4}{3}\pi r^3 \times 5.51$ and $dm/dr = 4\pi r^2 \times 5.51$, and it is immediately found that $U = f 16\pi^2 \times (5.51)^2 \times R^5/15 = 0.6 \ MgR$, which is $f \times 3.35 \times 10^{46}$ ergs $= 22.4 \times 10^{38}$ ergs.

For the actual earth, using the density distribution of Bullen's model A, a numerical integration gives the value 25.0×10^{38} ergs.

It is easy to calculate the elastic energy contained in the earth since we know as a function of the depth the pressure, the density, and the speeds of seismic waves, which are given by $\sqrt{(\lambda + 2\mu)/\rho}$ and $\sqrt{\mu/\rho}$. These yield the values of the elastic coefficients λ and μ. The elastic energy per unit volume is given by

$$\frac{3P^2}{2(3\lambda + 2\mu)} \tag{8-7}$$

The numerical integration (using the densities of Bullen's model A and the seismic velocities of Gutenberg and Richter) gives 7.88×10^{37} ergs (4.45 in the core and 3.43 in the mantle), which is 3.2 percent of the potential energy dissipated in the condensation of the earth and is thus not at all negligible.

To give an idea of the significance of these quantities, remember that the mass of the earth is 5.976×10^{27} g. If the average specific heat was of the same magnitude as that of rocks, which is 0.185, the energy dissipated in coming from infinity less the elastic energy, $(25 - 0.79) \times 10^{38}$, would correspond to a temperature rise of 54,000°C!

It is known that for a mass of the magnitude of the sun, condensation can release sufficient energy to raise the entire mass to a temperature of several tens of millions of degrees, which allows the beginnings of the nuclear reactions which furnish the energy radiated by the sun and the other stars.

To return to the earth, the transition from a state with a uniform density to the present state, having a dense core, would release 2.6×10^{38} ergs, which is 1040 cal/g and which is enough energy to raise the entire mass of the earth to 5000°C if the heat capacity was of the order of that of superficial rocks (this is without taking into account a possible increase of elastic energy made possible by the increase in pressure which took place as more dense material sank toward the center of the earth).

The genesis of the earth is often imagined as the condensation of a cloud of dispersed primitive matter. But it is not certain that this condensation happened all at once, which would have heated the entire mass of the earth as we have just seen. It is more plausible to imagine a successive accretion in layers, slow enough to dissipate the kinetic energy of rotation (by bodily tides or the action of a magnetic field?). The heat caused by condensation would then be released principally on the surface, and it could be imagined that it would at least be partly dissipated by radiation as the process continued. It is thus difficult to estimate exactly what was the heating of the terrestrial sphere as a result of initial condensation. But it seems certain that there was an initial heating.

In any case, the transition from a homogenous density distribution to the present state with a highly differentiated core required internal displacements of matter. The energy so dissipated must have heated the interior of the earth.

It is not at all certain that this potential energy dissipation had to occur at the time of the earth's genesis. It is perfectly possible that a certain differentiation is still taking place. A tenth of the energy dissipated in the transition from a sphere with a uniform density to the present globe with a differentiated core would have sufficed to supply the present heat flow (1.95×10^{20} cal/yr) for 2.9 billion yr without taking radioactive decay into consideration. *A fortiori*, the energy dissipated in mechanical forms (earthquakes, plate tectonics) only represents a very small proportion (on the order of a few parts per thousand) of the mechanical energy furnished by the differentiation of the core.

8-5 What conclusions can be drawn from these data?

We have tried to present the facts relative to different times in a thermal balance of the earth, considered in its entirety, independently of the ideas and hypotheses concerning its internal constitution so as to specify certain orders of magnitude and eliminate uncertainties which prevent any conclusions. It is clear that the thermal balance and the temperature distribution that results from it, presently or in the past, can only be established in the outline of a model incorporating information furnished by other disciplines such as seismology, gravimetry, magnetism, petrology, and geology in general, and coordinating this information in the framework of adequate hypotheses.

A model could be constructed to describe the present state of the earth: what we have seen of the time scale of all the phenomena of conduction on an earth-scale basis and the variations of the heat released by radioactivity make it apparent that such a model will only furnish valid results in the thermal domain when it is completed to describe the evolution of the earth in the course of 4.6 billion yr (which we now know represents its age). The present thermal system is directly dependent on this prior evolution, and, in particular, it would not be possible to analyze it without making hypotheses about the genesis of the earth.[1]

Obviously the scope of this book does not permit us to analyze the geophysical data upon which the current hypotheses on the internal constitution of the earth rely.[2] Thus, we will suppose that they are known and will only call on them occasionally to justify a choice between different hypotheses.

8-6 Outline of an energy balance

The heat flow is not the only form in which energy of internal origin is manifested. Without mentioning volcanoes, which appear essentially as thermal phenomena, the most spectacular manifestation of internal energy is folding and, in a more general way, the deformation of

[1]As an example of such a historical synthesis, coherent and implying a certain number of hypotheses, see: E. A. Lubimova, Thermal History of the Earth, in P. J. Hart (ed.), "The Earth's Crust and Upper Mantle," geophysical monograph 13, pp. 63–77, American Geophysical Union, 1969; and E. A. Lubimova, Theory of the Thermal State of the Earth's Mantle, in T. F. Gaskell (ed.), "The Earth's Mantle," pp. 232–323, Academic, New York, 1967.

[2]J. Coulomb and G. Jobert, "The Physical Constitution of the Earth," Oliver & Boyd, London, 1963. Also see Geophysique in J. Goguel (ed.), "Encyclopedie de la Pleiade," 1971; J. Coulomb and G. Jobert (eds.), "Traite de géophysique interne," vol. 1, Masson, Paris, 1973 (volume 2 in press); J. Goguel, Terre (constitution interne), Encyclopaedia Universalis, Paris, vol. XV, 1974, pp. 967–973; B. Gutenberg (ed.), "Internal Constitution of the Earth," Dover, New York, 1951.

rocks and the upheavals which form mountains. It is now known that this is a manifestation restricted to the continents; in general the plates which make up the crust (or more exactly, the lithosphere) glide with different speeds which are expressed by specific structures at the junctions between plates.

Let us at least try to estimate the order of magnitude for the energy dissipated by these different phenomena. One of the manifestations of present tectonic activity is the existence of earthquakes. Thanks to seismographs we have today a complete listing of all earthquakes above a certain magnitude. The estimation of the energy they release is difficult. It requires as an intermediary an empirical definition of magnitude, and then energy estimates for some chosen earthquakes, which leaves room for considerable uncertainty. It is generally conceded that from one magnitude to the next, the energy changes in a ratio of about 63 and the frequency in a ratio of 10. It is clear that the larger earthquakes represent the most important contribution to the total energy released. The figure proposed, 10^{25} ergs/yr, is equivalent to 3×10^7 kW $\cdot$ yr/yr, noting that the major portion of this energy is dissipated in a few seconds.

Outside of seismic energy, for which we will suppose that the present rate can be taken as representative of past geological ages, tectonic energy is partially manifested in the creation of relief and partly by the deformation, sometimes very strong, of rocks that are mechanically resistant.

It is easy to calculate the potential energy of the visible relief using the statistics of surfaces classified by altitude, which is expressed by the hypsographic curve. The potential energy of continental relief is calculated separately with respect to sea level and gives 4.7×10^{21} kJ, and the potential energy of relief at a constant oceanic depth is 29×10^{21} kJ. But we must take isostatic compensation into account. To all visible relief there correspond density deficits situated at a certain depth that make us underestimate the previous figures in the ratio of the density of the crust to the variation of the density which determines the compensation, commonly taken as 0.6 g/cm^3. This gives $2.67/0.6 = 4.5$ for the continents and $1.67/0.6 = 2.7$ for the oceans, to account for the Archimedean pressure. In reality, these factors are quite poorly determined, the subterranean densities being hypothetical, and could be in error by from 10 to 20 percent.

Geology shows us that ancient relief was destroyed by erosion and that all the relief that we now observe resulted from relatively recent upheavals (i.e., post-Miocene), even when the material raised had been deformed at a much earlier time. We will suppose then that the rate of energy dissipation in the formation of relief corresponds to the creation in 10 million yr of the continental relief equivalent to that now existing. The oceanic relief is in totality much older, and its creation should be spread out over a duration of 200 million yr or perhaps more. For the continental relief, $4.7 \times 4.5 \times 10^{21}$ $kJ/(10^7 \times 3.15 \times 10^7) = 6.7 \times 10^7$ kW. The creation of oceanic relief in 200 million yr would absorb $(29 - 4.7) \times 2.7 \times 10^{21}/(2 \times 10^8 \times 3.15 \times 10^7) = 1.07 \times 10^7$ kW, which is in total about 8×10^7 kW.

It is very difficult to estimate the energy absorbed by rock deformations because we only know that the stresses under which they were produced (for very slow processes which essentially result from recrystallization) were certainly much less than that which we observe in the laboratory for rapid deformations by different processes. We know that the total energy absorbed is divided between deformation and creation of relief in such a way that the total energy absorbed, for a given deformation, is a minimum and must be reflected in comparable orders of magnitude for these two terms.

Plate theory implies that over the entire surface of the globe rigid plates formed by the lithosphere glide with respect to the internal regions of the earth by means of the viscosity of a region called the *asthenosphere*, with a speed of a few centimeters per year. The best indication that we have for the viscosity of the asthenosphere is furnished by postglacial readjustments in Scandinavia. For a surface area whose diameter is 2,000 km, half of the isostatic disequilibrium has vanished in 6,000 yr and this delay can hardly be attributed to anything else than a viscous action.

If the depth of the asthenosphere is great, a viscosity of the order of 10^{21} poise would be deduced. If the asthenosphere has only a limited depth, we would naturally find smaller values, inversely proportional overall to the cube of the thickness, which would be of the order of 10^{19} poise for a thickness of 100 km.

Let v be the speed of the plate in centimeters per year. We will set the thickness of the asthenosphere as $h \times 100$ km, which gives a value of the shear stress, produced by the coefficient of viscosity times the gradient of the speed, of $\eta \times (v/h) \times 3.15 \times 10^{-14}$ dyn/cm^2 and of the work in a year of $(\eta v^2/h) \times 3.15 \times 10^{-15}$ erg/cm^2. If v is the root-mean-square plate speed for the entire earth, the total power is $(\eta v^2/h) \times 5 \times 10^{-14}$ kW. For $h = 1$, $v = 3$ cm/yr, and $\eta = 10^{19}$, we find 4.5×10^6 kW, and for $h \times 3$, $\eta = 10^{21}$, and $v = 2$ cm/yr, we find 6.6×10^7 kW. We would find a power of the same order or smaller if we supposed a thinner asthenosphere with a lower viscosity. It is very probable that this viscosity varies greatly with depth, but it would be of little use to introduce in the model a viscosity law of which we know nothing.

From the collection of these estimates, rough as they are, we can conclude that the mechanical energy spent in tectonic processes must be of the order of 1 or 2×10^8 kW, which is very small compared to the geothermal flux of 2.56×10^{10} kW. It is thus very likely that we should look for a mechanism by which this mechanical energy can be derived from thermal processes without, however, necessarily excluding a direct transformation of gravitational potential energy caused by differentiation of the core, or as a consequence of other density differentiations in the midst of the mantle, to mechanical energy.

It is easy to show that the hypothesis of a contraction caused by the cooling of the earth, which was suggested at the beginning of last century, would not be satisfactory. To show this, it suffices to calculate the mechanical energy accumulated in elastic form resulting from the stresses due to cooling. These stresses increase in proportion to the time, and thus the elastic energy increases as the time squared. The average power available is proportional to the time that the stresses accumulated. But the calculation done[1] for the most rapid cooling imaginable, which is that of the model of Buffon, without considering radioactive effects, shows that these stresses would have had to accumulate for a time period greater than the age of the earth and would reach values of hundreds of kilobars, which no rock could withstand without rapidly deforming. Thus, we can formally eliminate this old hypothesis of contraction.

8-7 Convection

The only mechanism left which we can invoke is that of convection currents. But we should be ready to recognize convection in unusual forms. For example, the eruption of a volcano, in which the weight of the overburden, consisting principally of consolidated lava, exerts pressure on the lava reservoir which allows lighter molten lava to rise to the surface, can be considered as a form of convection (see Fig. 8-2). If the molten lava, before solidifying, released volatile substances which contributed to lightening the lava and causing it to rise, a process of gravitational differentiation would be added to the thermal convection.

Convection implies a uniform chemical composition (with the reservation that differentiations which could occur in the course of convection could contribute to activate the gravitational potential energy). The increase in density with depth between 400 and 900 km in the earth, which seems to be too great to result from compression of matter alone, is attributed to the appearance of high-pressure mineral phases. The best known is the spinel form to which olivine can invert above a pressure which increases with the magnesium concentration. But other transformations are also possible; silica can appear in the form of

[1]Jean Goguel, Une estimation de l'énergie mécanique disponible dans la contraction par refroidissement, *Tectonophysics*, vol. 2, no. 5, pp. 395–400, 1965.
Jean Goguel, La contraction thermique peut-elle expliquer les déformations tectoniques?, *Bull. Volcanologique*, vol. XXXIII, no. 1, pp. 89–100, Napoli, 1969.

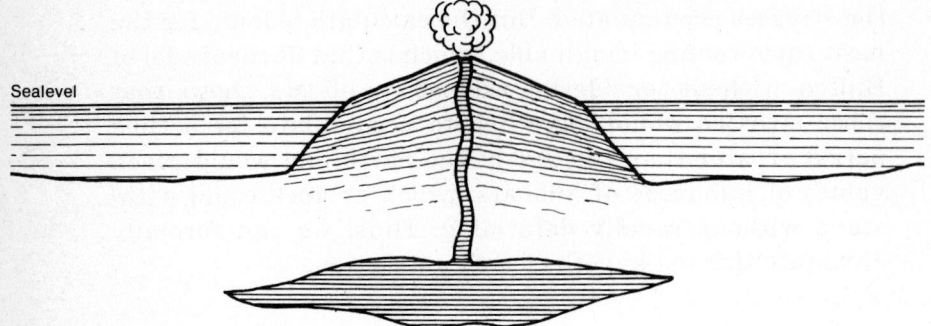

Sealevel

FIGURE 8-2   Sketch of a Hawaiian volcano showing how a lava eruption
             can be considered as a type of convection.

stishovite above 100 kbars, with a density at atmospheric
pressure, greater than 4.

Convection also implies a certain fluidity, but we shall
see that the viscosity can be very high; but it is necessary
that weak stresses can cause a very slow deformation
without it being necessary for the stress to exceed a threshold of plasticity.

Convection can only be produced if the thermal gradient is greater than or equal to the adiabatic gradient
which corresponds to the temperature variation due to the
change of pressure with depth. As best we can guess from
mineral properties studied in the laboratory, this gradient
may be of the order of 0.25 to 0.30°C/km.

From the moment that a thermal gradient exists, it
gives rise to a heat flux by conduction which does not interfere with convection. The latter implies a heat transport by
moving matter, and it is to this heat transport that Carnot's
principle can be applied in order to calculate the maximum
proportion of heat transformed into mechanical energy as a
function of the temperatures at the base and top of the convective system. This proportion will be given by the ratio of
the temperature difference between the base and the top
and the absolute temperature of the base. If the height is
limited, with a gradient nearly that of the adiabatic gradient, this efficiency must be very low.

But the viscosity in the midst of this region is certainly high, and part of the mechanical energy produced is dissipated by viscosity in the movements of the currents. It could even happen (which is what we observe in many convective movements which do not have an external effect) that the total mechanical energy produced is dissipated by viscosity; in general, this will be the case for a substantial amount of the energy.

As a first approximation, we can give a description of a system of convection currents relative to the average situation, which has a regular thermal gradient. For any sort of model of currents perturbing this regular gradient, we will consider a distribution of the thermal perturbation, through a factor $\Delta\theta$, and a distribution of velocities, also through a factor, that will be characterized by a typical velocity V. Let L be the characteristic dimension which defines the scale of the motion. The velocity gradients are proportional to V/L; the stresses which result from these gradients go as $\eta V/L$, the power dissipated per unit volume as $\eta V^2/L^2$, and the total dissipated power goes as $\eta V^2 L$. In addition, the fluctuations in density are proportional to $\lambda\,\Delta\theta$ if λ is the coefficient of thermal expansion and the pressure variations go as $\lambda L\,\Delta\theta$. The mechanical power furnished is equal to an integral extended over the product of the velocity times the fluctuation in the pressure gradient with respect to equilibrium, an expression which must vary as $L^3 V \lambda \Delta\theta$.

In order that this system of currents can act under the conditions we are considering, it is necessary that $\lambda\Delta\theta VL^3/(\eta V^2 L)$ exceed a certain numerical value which depends on the convection model and on the definition of the velocities, the dimensions, and the variations of the characteristic temperature. This ratio is equal to $\lambda L^2\,\Delta\theta/\eta V$, and it is always possible to take V small enough, whatever the viscosity, so that this condition is fulfilled. But the dissipated mechanical power is given by an expression proportional to $L^3\lambda V\,\Delta\theta$; $\Delta\theta$ will not take too large of a value and will not exceed the temperature difference between the base and the top of the system, which limits the mechanical power which can be produced. It is likely that this limitation acts and is lower than the limitation resulting from Carnot's principle.

The preceding considerations suppose nothing about the form of the convection currents or the model which represents these currents and which can—as we have indicated at the beginning of this section by an example—take the most diverse forms. It is not possible to undertake a calculation of such models because their particularities can result from quite complex phenomena such as viscosity variations with the temperature.

In fact, there exist no models of convection currents at present which can be put to an exact calculation even by simplifying the representation of physical properties (viscosity taken as uniform and independent of the temperature, etc.). There are models which lend themselves to calculations concerning the initial stage, for which, supposing a thermal perturbation, we can investigate whether the movements of the fluid will permit the perturbation to amplify or disappear by conduction more rapidly than the movements caused by density differences can augment it. This is a stability calculation analogous to the calculation we did in Chap. 3 to analyze the possibility of a convection of a very different nature (in a porous rock). We have no valid mathematical model prepared to account for convection in full development, assuring an effective heat transport, furnishing mechanical power, and absorption of part of that power by viscous friction.

Now, given the dimensions of the earth and the extremely high viscosity of the mantle, it is indispensable to give an idea of the orders of magnitudes of the currents which we are considering. We can derive them by approximate reasoning relying upon very rough models.

The simplest way to represent a linear convection cell of width $2a$, depth $2b$, and length c, is to write the speeds of the fluid:

$$U = v_0 \frac{a}{b} \cos \frac{\pi x}{2a} \sin \frac{\pi y}{2b}$$

$$V = -v_0 \sin \frac{\pi x}{2a} \cos \frac{\pi y}{2b}$$

The temperature in an ascending or descending current must be characterized by its difference from the adiabatic profile (it would be more correct to characterize the current by its entropy, which is conserved). If G is the superadiabatic gradient (excess of the actual gradient over the adiabatic gradient), the margin of possible variations in temperature is $2Gb$ and the difference between temperatures in the currents and the average is $Gb \sin(\pi x/2a)$. The variation of the density due to thermal expansion is $\rho g \lambda \, \Delta\theta$, and we will take $\rho = 3, g = 900, \lambda = 5$ to 8×10^{-6}, and for the viscosity, $\eta = 10^{21}$.

It is easily found that the power dissipated by viscosity from a cell is

$$W = \frac{2ac}{b} \, \eta v_0^2 \pi^2 \left(\frac{a^2 + b^2}{4ab}\right)^2 \tag{8-8}$$

The motive power results from the vertical displacement of the fluid occurring in a pressure gradient modified by the variation of density, and it is found to be

$$P = 2ac \left(\frac{2}{\pi}\right) g\rho\lambda G v_0 b^2 \tag{8-9}$$

To get orders of magnitude applicable to the earth from these formulas, we will express the vertical velocity V in centimeters per year ($V = v_0 \times 3.15 \times 10^{-8}$) and the height of the convective system, $2h$ in kilometers ($h = b \times 10^{-5}$), and we will extend the calculation over the entire earth's surface by taking an average radius for the system of 5,000 km, which gives a surface area of $\pi \times 10^{18}$ (which replaces $2ac$). We will express the superadiabatic gradient S in degrees Celsius per kilometer, which means $S = G \times 10^5$.

From these substitutions it is found that $P = 10^4 \, SVh^2$ kW and $W = 10^{10} \, V^2/h$ kW, and the heat flux transported by convection over and above the flux transported by conduction equals $2.7 \times 10^{10} \, hV$ cal/s or $10^8 \, hV$ kW, an expression which could take excessively large values for some models which have been proposed. Thus, it is the heat flux to be transmitted which will limit the temperature differences of the convective current to smaller values than we have allowed as the possible maximum (Gh); it may also happen that the velocities are smaller or that the currents instead of distributing themselves over the entire available surface, as our model supposed, are much more localized.

From the preceding formulas the condition can be given that

$$V < S \left(\frac{h}{100}\right)^3 \qquad (8\text{-}10)$$

If $S = 1$ (it can scarcely be imagined that S will exceed a value of 2 or 3), this gives a velocity limit $V = 1$ cm/yr for a system 200 km high and $V = 1$ m/yr for a system 1,000 km high.

These formulas show us how the power furnished depends on the height of the system. Suppose that the velocity is half of the limit and also that $S = 2$. We will then have $V = (h/100)^3$ and the power furnished will be given by $P = 2 \times 10^{-2} h^5$ kW, of which half will be absorbed by the viscosity and the other half will be utilizable.

Note especially the exponent of h in Eq. (8-10); the mechanical power which can be produced by convection, appearing in the mantle where the thermal gradient is determined in another way (very likely this gradient is close to the melting-point gradient which is produced by pressure), increases enormously with the height of the convective system.

Our calculations are much too rough for us to obtain the value of the power furnished; Carnot's principle imposes a much surer limit. But it should be remembered that a convective system with a small height, of the order $2h = 200$ km, will furnish insufficient mechanical power and that a thicker system, of the order of 1,000 km, seems much more probable. The maximal velocities of the currents could then be of the order of a meter per year.

On the whole, it appears that we can retain the hypothesis of convection currents, originating from a thermal flux in the mantle and producing mechanical energy which is manifested in internal activity of the earth. But the model utilized for these calculations should absolutely not be considered as a description; actual convective currents can be quite different.

Remember that the viscosity very likely changes with depth, and the action of dissipating energy through viscous friction by currents locally raises the temperature and reduces the viscosity; it thus appears that the current forms are certainly very different from the forms discussed previously. In particular, ascending currents, hotter and less viscous, tend to be more rapid, thus occupying a small cross-sectional area, and are separated from larger descending currents by a slippage surface along which the temperature will increase sufficiently to greatly reduce the viscosity, rather than by a zone with a regular velocity gradient.

We must emphasize a difference between the convection currents that we have the opportunity to observe and those currents which can exist in the interior of the earth. In the first type, inertia plays an appreciable role, and, as a consequence, there is a tendency for the current system to maintain the same sort of motion. It is not the same for currents in the mantle, where the kinetic energy, with velocities of the order of magnitude previously indicated, is totally negligible. For harmonic models which lend themselves to calculation, it can be shown that the kinetic energy corresponds to the energy absorbed by viscous friction in a time which is of the order of $\frac{1}{100}$ s. As a consequence, nothing excludes the possibility that the movement is very irregular in time with alternations of halts (when elastic energy is accumulated) and more rapid movements.

Finally, let us remember that all the considerations on the power furnished by any system of thermal convection currents can be false if there is at the same time a differentiation which is expressed by a density difference (at equal temperatures) between ascending and descending currents. A lighter residue then will accumulate at the top of the system and a heavier residue at the bottom. Such a progressive differentiation could have continued since the beginning of geological time and furnished a considerable part of the mechanical energy dissipated in tectonic events.

These theoretical considerations show that the hypothesis of convection currents does not run into any objections (at least if they exist over a sufficient height). But these considerations do not give us any valid indication of what their form or their depth could be.

What indications of these characteristics can we obtain by other methods? There is a decline in seismic velocities at a depth of around 150 km which has been interpreted as indicating the beginning of fusion and which marks the top of the asthenosphere. This interpretation is occasionally invoked to allow plate gliding and isostatic readjustments. It is natural to assume that the upper limit of convection currents also occurs there. What is their base? Models have been proposed by different authors attributing a quite small height for these currents. This seems unlikely given the difficulty that there would be in producing the energy dissipated in tectonic processes. It is poorly understood how heat transport by conduction alone could exist beneath convection currents without producing thermal gradients leading to other zones of incipient fusion. It seems that it is much more likely that these currents occur over a large depth in the mantle, and there are no peremptory arguments to dispute such a hypothesis.

The classical mechanism proposed for plate tectonics is in fact a form of convection, in particular the ascent of intrusive rocks, basic or ultrabasic, in fissures which mark the openings of rifts. It is known that the magnetization acquired in the earth's field in prior times by rocks in the course of their cooling is the origin of magnetic anomalies parallel to the rift on both sides. Actually, the geothermal flux is greater than normal (up to eight times) on the axis of the rift and up to 100 km on either side. The additional amount of flux by comparison with the normal value can be of the order of 3×10^9 cal/cm$^2 \cdot$ yr, which for a supplementary annual thickness of 2 cm of intrusive rock effectively represents the heat liberated by a cooling of 500°C over 60 km of depth. But it is quite evident that such an average balance would not completely describe the phenomenon, and all models proposed to describe plate action include a geothermal interpretation, destined to interpret the diffusion of heat away from the injections.

The descending convective branch in the lithosphere and even lower would be represented by the subduction of plates under a Pacific-type island arc (or under a continental plate like South America). But here, the geothermal flux distribution, with higher than average values of about 2 $\mu cal/cm^2 \cdot s$ behind the arc, is not so easily interpreted; the flux distribution makes it necessary to invoke secondary phenomena (refusion, secondary cells of convection, etc.) which remain very hypothetical. But there again, the geothermal interpretation is essential for all proposed models of plate action. However, the detailed analysis of such models takes us away from our subject, and they can be found in descriptions of plate theory.

The arrangement of convection currents in the asthenosphere is probably related to the behavior of these convective phenomena in the plates themselves, and this must be reflected in their general disposition, with ascending currents in the vicinity of mid-oceanic rifts and descending currents in subduction zones and along mountain ranges which resulted from compression. But it can well be thought that certain geometric constraints which appear in the case of plates as transform faults which offset the rifts do not act on the currents whose axes can take much more supple forms. Are there other areas of ascending and descending currents which should be held responsible for the evolution of certain regions in the midst of plates such as uplifts (followed by erosion) which would result from weak compression or depressions which permitted the accumulation of thick sedimentary basins and which could result from a small extension? I can only ask the question.

It has been pointed out recently that some volcanic sequences appear more and more ancient as the distance from the region of presently active volcanoes increases (the best example is furnished by the Sandwich Islands). These sequences appear more and more ancient as the distance of a plate above a fixed heat source which has been identified as a plume of hot material escaping from the lower

mantle at points which have remained stationary for durations approaching hundreds of millions of years. Although
the depth from which these plumes issue remains very
poorly determined, it is a first indication of what could be
the arrangement of currents in the lower mantle. But of
course, there could not be any ejection of material by a certain number of quasi point sources unless it was compensated by the descent of cooler material, perhaps over a much
greater surface area and with a very small velocity. It would
be geometrically improbable that the descent was produced
radially in every direction in a sphere. It can be imagined
that the system of ascending currents has the form of a
network but that the upward motion tends to reach greater
and greater velocities and be concentrated at the intersec-
tions of the network. This arrangement is what would be
expected of matter so completely solidified that the viscosity is very high except where the dissipation of energy by
viscous friction raises the temperature.

The existence of convection currents in the lower
mantle is made very probable by an interpretation which is
given today to the origins of the internal magnetic field. It is
thought to originate in the core by magnetohydrodynamic
processes. These processes imply, as a source of energy, convection currents of thermal origin stirring the molten iron
which constitutes the core with velocities much higher than
in the mantle, of the order of 100 km/yr. It can be shown that
for such speeds, the Coriolis force caused by terrestrial rotation is not negligible and tends to make circulations predom-
inate in planes perpendicular to the axis of rotation. This
yields the N-S component of the average magnetic field. The
displacement of conductive iron in the magnetic field gives
rise to electric currents which, in their turn, maintain the
magnetic field.

One may wish to know the amount of energy spent in
the creation and modifications of the terrestrial magnetic
field, but this can only be estimated and that only by
default. However, let us indicate the elements of the
argument. If the portion of the magnetic field which
represented the effect of a central dipole, and which consti-

tutes the essential portion of the field observable at the surface, was caused by electric currents circulating on the surface of the core, it would correspond to an energy of 6.3×10^{25} ergs, which is 2×10^8 kW · yr. But if the currents which produced the field circulated in a sphere of lesser radius, this energy would increase as the inverse of the radius cubed—and there are many reasons to think that the currents are in fact distributed in the volume of the core. Thus, we are led in the calculation of the energy of the magnetic field, to consider successive terms of its harmonic development, although this development is a purely mathematical artifice; these terms are simply added. Now, it is observed that the energy corresponding to each of the terms of order 4, 5, 6, 7, and 8 would be of the same order as the energy of the first term (central dipole) if the currents which caused them circulated in a sphere of radius 2,650 km, which is three-quarters (77 percent) of the radius of the core. Our observations, necessarily confined to the earth's surface, do not inform us about higher terms, but it is perfectly possible that they also correspond to energies of the same order and represent local irregularities of the field of the core. The total energy of the field, then, at a minimum is 10 to 20 times greater than the number previously cited, and could even be greater. Theoretically, the maintenance of a magnetic field does not absorb any energy, but its creation and modification does. At the surface we observe variations in the magnetic field with periods of about 1,000 yr. But before deducing an order of a magnitude for the energy absorbed from this period, it is necessary to remark that if the activity in the core implied more rapid variations in the magnetic field, they would not be observable at the surface because they would be compensated by induced currents in the conducting lower mantle. It is this filtering in time which limits the observable rate of change at the surface, and thus much more rapid variations of the field in the core are not at all excluded. In total, the facts in our possession risk leading us to a grave underestimate of the power absorbed in changes of the magnetic field, which would be at least 10^7 to 10^8 kW.

It would also be necessary to take into account, besides

the power dissipated by the joule effect, the power dissipated in electric currents in magnetohydrodynamic phenomena, and it is likely that the thermodynamic efficiency of the convection currents in the core, which allow these phenomena, must be very low. So much so that the heat flux generated in the core must be much higher than the above figures.

It is very unlikely that such flux can be transmitted by conduction through the base of the mantle without leading at least to partial fusion. It seems that the most likely solution is to admit that convection currents occur over the entire depth of the mantle.

8-8 A hypothetical model for the mantle

The extrapolation of properties measured at the surface lead us to think that the melting-point gradient of the ultrabasic silicates which must constitute the mantle (but what is the gradient for high-pressure mineralogic forms?) is distinctly higher than the adiabatic gradient. From this it can be concluded that if the convection currents maintain a gradient nearly that of the adiabatic, then the mantle must have begun to crystallize at its base. But in reality, near the fusion point the adiabatic thermal profile becomes entangled with the thermal profile of solidification. In an adiabatic transformation, it will be the proportion of solid and liquid phases which will vary, the solid phase increasing toward the bottom. This is not incompatible with what is known of the form of convection currents in the lower mantle.

From the preceding, we can imagine that in the mantle the temperature everywhere would be close to the temperature corresponding to the beginning of fusion, giving the mantle the ability to deform with perhaps a very high viscosity. The fusion conditions of ultrabasic rocks are not sufficiently well understood at high pressures, where minerals take their high-pressure forms, to permit us to calculate the temperatures; at least we can say that this depends only on determinations which are in principle possible in the laboratory.

Such a distribution of temperatures is certainly not a simple coincidence. First, let us imagine an earth that at its beginnings was heated sufficiently so that all matter was molten. Differentiation of the core could then easily occur with a complementary production of heat. As soon as radiation had sufficiently cooled the surface so that a solid crust could form, the conductive transfer through the crust would become very slow, in spite of the relatively high thermal gradient existing there. On the other hand, convection would continue to occur in the mantle beneath and assure easy heat transfer to the base of the crust. The interior of the earth could then cool until the moment when the beginning of solidification rapidly increased the viscosity. But this solidification could not run to its ultimate end because the latent heat of solidification would not be removed by conduction through the solid rock, given the low value of the gradient inherited from the prior convective system.

If, on the contrary, the earth was formed cool, it would progressively heat up by radioactivity and compression and we would have to concede that this heating had reached, at least in some places, the fusion temperature. Then convection would start which would allow removal of heat much more easily than by conduction alone and would limit the temperature increase in the molten region and regularize the temperature distribution there. If the entire mantle had thus reached the fusion temperature, the analogy with the preceding imagined situation of an earth formed at a high temperature would be complete. If it is conceded that the mantle only reached the fusion temperature in a few places over a limited thickness, the fusion temperature would permit the establishment of a convective system and the differentiation of the core would remain to be explained. This raises enough difficulties so that it seems to us that this hypothesis must be discarded.

Recall that the model we imagined would not have to be stationary; the decrease of the percentage of radioactive substances in the crust would necessarily lead to a temperature decline at its base which is inevitably reflected by conditions at the top of the mantle. It is perhaps necessary to admit that in the lower Precambrian the convective system occurred over the entire depth of the mantle, thus assuring its homogeneity. In contrast the crust became chemically differentiated and lighter. We have already noted that the geothermal flux and the gradient ought to have been higher then, which would have important geological consequences. In conclusion (see Fig. 8-1), from the upper Precambrian, the top of the mantle must have cooled, solidified, escaped convective action, and made a mechanical connection with the crust to form the lithosphere. It can be asked in what way such an evolution would be translated into the tectonic, magmatogenic, and metamorphic characteristics of mountain ranges of different ages compared with those of the lower Precambrian.

appendix

Table A-1
Useful functions

x	e^{-x^2}	$\mathbf{erf}(x) = \dfrac{2}{\sqrt{\pi}} \displaystyle\int_0^x e^{-y^2}\,dy$	$\mathbf{erfc}(x) = 1 - \mathbf{erf}(x)$	$\Phi(x) = \sqrt{\pi}\displaystyle\int_x^{\infty}\mathbf{erfc}(y)dy = e^{-x^2} - \sqrt{\pi}x\,\mathbf{erfc}(x)$	
				$\Phi(x)$	$\Phi(-x)$
0	1	0	1	1	
0.1	0.99005	0.1125	0.8875	0.8317	1.1862
0.2	0.96079	0.2227	0.7773	0.7022	1.3932
0.3	0.91393	0.3286	0.6714	0.5559	1.6194
0.4	0.85214	0.4284	0.5716	0.4942	1.8639
0.5	0.77880	0.5205	0.4795	0.3521	2.1345
0.6	0.69768	0.6038	0.3962	0.2756	2.4025
0.7	0.61263	0.6778	0.3222	0.2122	2.6936
0.8	0.52729	0.7421	0.2579	0.1611	2.9970
0.9	0.44486	0.7969	0.2031	0.1204	3.3108
1	0.36788	0.8427	0.1573	0.08871	3.6336
1.2	0.23693	0.9103	0.0897	0.04591	4.2997
1.4	0.14086	0.9523	0.0477	0.02235	4.9851
1.6	0.07730	0.9763	0.0237	0.01002	5.6818
1.8	0.03916	0.9891	0.0109	0.004355	6.3852
2	0.01831	0.9953	0.004677	0.001735	7.0915
2.5	0.00193	0.9996	0.000407	0.0003278	8.8626
3	0.000123	0.9999779	0.0000221	0.000008063	10.6347

Table A-2
Properties of water and of saturated steam

Temperature, °C	Pressure, bars	Latent heat, cal/g	Specific volume, cm³/g	
			Water	Steam
100	1.013	539.7	1.0434	1,673.
120	1.9854	526.6	1.0603	891.7
140	3.6136	512.5	1.0798	508.66
160	6.180	497.1	1.1021	306.85
180	10.027	480.2	1.1275	193.85
200	15.550	461.6	1.1565	127.19
220	23.202	441.35	1.1900	86.062
240	33.48	419.45	1.2291	59.674
260	49.941	397.2	1.2755	42.149
280	64.19	368.6	1.3321	30.133
300	85.92	335.6	1.4036	21.643
320	112.89	295.7	1.4992	15.451
340	146.08	245.5	1.639	10.779
360	186.74	172	1.894	6.943
374.15	221.29	0	3.1	

index

index